Integrated Series in Information Systems

Volume 26

Series Editors

Ramesh Sharda
Oklahoma State University, Stillwater, OK, USA

Stefan Voß
University of Hamburg, Hamburg, Germany

For other titles published in this series; go to
http://www.springer.com/series/6157

Robert P. Schumaker • Osama K. Solieman
Hsinchun Chen

Sports Data Mining

Springer

Robert P. Schumaker
Cleveland State University
Cleveland, OH 44115, USA
rob.schumaker@gmail.com

Osama K. Solieman
Tucson, AZ 85704, USA
osolieman@gmail.com

Hsinchun Chen
University of Arizona
Tucson, AZ 85721, USA
hchen@eller.arizona.edu

ISSN 1571-0270
ISBN 978-1-4614-2691-2 ISBN 978-1-4419-6730-5 (eBook)
DOI 10.1007/978-1-4419-6730-5
Springer New York Heidelberg Dordrecht London

Printed on acid-free paper

Springer is part of Springer Science+Business Media (www.springer.com)

Preface

Aims

Sports data mining has experienced rapid growth in recent years. Beginning with fantasy league players and sporting enthusiasts seeking an edge in predictions, tools and techniques began to be developed to better measure both player and team performance. These new methods of performance measurement are starting to get the attention of major sports franchises including baseball's Boston Red Sox and Oakland Athletics as well as soccer's AC Milan.

Before the advent of data mining, sports organizations relied almost exclusively on human expertise. It was believed that domain experts (coaches, managers and scouts) could effectively convert their collected data into usable knowledge. As the different types of data collected grew in scope, these organizations sought to find more practical methods to make sense of what they had. This led first to the addition of in-house statisticians to create better measures of performance and better decision-making criteria. The second step was to find more practical methods to extract valuable knowledge using data mining techniques. Sports organizations are sitting on a wealth of data and need ways to harness it.

This monograph will highlight current measurement inadequacies and showcase techniques to make better usage of collected data. Properly leveraging sports data mining techniques can result in better team performance by matching players to certain situations, identifying individual player contribution, evaluating the tendencies of opposition, and exploiting any weaknesses.

Audience

The primary audience for this monograph includes the following:

- *IT Academic Audience:* College professors, research scientists, graduate students, and select undergraduate juniors and seniors in computer science, information management, information science and other related public policy disciplines who are interested in data mining and its applications in various emerging technology fields.

- *Sports Management Academic Audience:* College professors, research scientists, graduate students and select undergraduate juniors and seniors in sports education and management related fields who are interested in an overview of sports measurement techniques and application to the sports environment.
- *Sporting Industry Audience:* Executives, managers, analysts and researchers in the business of sports, research institutions that are actively conducting sports data mining research and industry analysts who are interested in identifying critical inventions and innovations that can lead to major commercial successes in the industry.
- *Sports Enthusiasts Audience:* Individuals and sports-related organizations that want to learn how to gain an edge using modern data mining tools and techniques to uncover hidden knowledge.

Future

Over the next several years, sports data mining practices will be faced with several challenges and obstacles. The most obvious of which is to overcome the years of resistance by the members of sporting organizations that would rather stick with a traditional way of doing things. Aside from the challenges that are faced, sports data mining currently sits at a pivotal junction in history with many opportunities just waiting to be grabbed. Some avenues of opportunity will be pursued quickly, while others may take years or decades to become fruitful. In any event, sports data mining today is still in its infancy. While some first steps were made with pioneers such as Dean Oliver and Bill James, the next few years will become a transition period as the technology begins to mature within the sporting community and become more commonplace. New metrics, algorithms and ways of thinking will begin circulating themselves as the field enters puberty and begins to mature. The coming decades will be fascinating to watch.

Robert P. Schumaker — Cleveland, OH
Osama K. Solieman — Tucson, AZ
Hsinchun Chen — Tucson, AZ

Contents

List of Figures

List of Tables

Chapter 1
Sports Data Mining: The Field

Chapter Overview

Incredible amounts of data exist across all domains of sports. This data can come in the form of individual player performance, coaching or managerial decisions, game-based events and/or how well the team functions together. The task is not how to collect the data, but what data should be collected and how to make the best use of it. By finding the right ways to make sense of data and turning it into actionable knowledge, sports organizations have the potential to secure a competitive advantage versus their peers. This knowledge seeking approach can be applied throughout the entire organization. From players improving their game-time performance using video analysis techniques, to scouts using statistical analysis and projection techniques to identify what talent will provide the biggest impact, data mining is quickly becoming an integral part of the sports decision making landscape where manager/coaches using machine learning and simulation techniques can find optimal strategies for an entire upcoming season.

The first part of the problem is to identify the metrics of performance. Many existing sports metrics can be easily misused or worse, do not measure performance in the context of scoring more points than their opponents, which is the ultimate goal of any sports organization. In later chapters we will discuss many of the problems with using these statistics as performance measures and introduce the reader to newer sport-specific statistics that take into account point scoring behavior in their performance assessments.

The second part of the problem is to find interesting patterns within the data. These patterns could include the trends and tendencies of opposing players/teams, determine the onset of injury through the monitoring of workout performances or make sport-related predictions based on historical data. In later chapters we will discuss different machine learning and simulation techniques that can be applied to many sports.

Professional sports organizations can be multi-million dollar enterprises with millions of dollars spent on a single decision. With this amount of capital at stake, just one bad or misguided decision has the potential of setting an organization back

R.P. Schumaker et al., *Sports Data Mining*, Integrated Series in Information Systems 26,
DOI 10.1007/978-1-4419-6730-5_1, © Springer Science+Business Media, LLC 2010

by several years. With such a large array of risks and a critical need to make good decisions, the sports industry is an attractive environment for data mining applications.

1 Definition

We postulate that there are a myriad of relationships between sports and the techniques designed to make use of the sport-related data. These relationships depend upon the nature of the organization and their need for historical data. We present our proposed hierarchy of sport and sport data relationships in Table 1.1.

The first relationship type is no relationship between sports and game-based data. These are simply professional organizations that play their game, record data generated by the players and do nothing further with the data. For these organizations, the collected data is simply a way of either recording the game events or is done as a result of tradition. Amateur sports organizations will typically exist at this level where the emphasis is on entertainment or learning the fundamentals (e.g., sports organizations with child participants or adult recreational sports clubs). The next type of relationship is to use human domain experts to make predictions based on experience. At one time it was believed that these experts (coaches, managers and scouts) could effectively synthesize their observations and experiences together, to make satisfactory decisions. The decisions generated from this level of relationship are typically based on hunches or instincts, not solid data. These decisions could include executing certain plays or making certain player substitutions because the decision feels "right" to the decision-maker, without regard to prior data. The third type of relationship is where domain experts start to make use of historical data. Decisions from this level would include making favorable player match-ups based on prior data and executing plays that have historically had better than average results. The fourth type of relationship begins to incorporate statistics into the decision-making process. These statistical measures could be simple frequency counts of particular events or more complex methods that divide team effort and assign each team player credit based upon their contribution towards achieving a goal. The measurement of player and team performance is used extensively at this level as are the creation of new methods of performance measurement. The statistics are used as a tool to augment the domain expert in their decision-making capacity. The fifth level of our postulated relationship

Table 1.1 Hierarchy of sport and sport data relationships

Level	Relationship
One	No relationship
Two	Human domain experts make predictions using instinct and gut feeling
Three	Human domain experts make predictions using historical data
Four	Use of statistics in the decision-making process
Five	Use of data mining in the decision-making process

between sports and sports-related data is the use of data mining techniques. These techniques differ from the previous level because data mining techniques have the potential to be generalized into new situations and predictions made from them. While statistical techniques still lay at the heart of data mining, the statistics are used to distinguish between a pattern of something interesting, such as a trend or tendency of an opposing player, versus random noise and allows researchers and sports organizations to test hypotheses and make predictions from the output (Piatetsky-Shapiro 2008). The statistics themselves do not explain the relationship; that is the purpose of data mining. This level of relationship also has the potential to be used to augment the decisions of domain experts or used independently to make decisions without the input of the domain expert. The latter usage of data mining techniques, without the influence of human input, could be argued to be free of many of the biases that permeate the human decision making process. An example of this would be an organizational scout falling in love with specific performance attributes of a player and ignoring the player's weaknesses. By removing the potential of human biases from the decision making process, coaches and managers have the potential to manage more effectively and make objective decisions that can help the organization.

Many professional organizations reside within the upper levels of our postulated sports data relationship where data is used to some extent in the decision-making process. Using professional baseball as an example, many teams could be argued to be level three or four depending upon the domain expert involved. These organizations use data to find appropriate player matchups (e.g., batters versus pitchers) as well as to indicate the likelihood of a particular play's success (e.g., the likelihood of successfully stealing a base against a specific catcher). Higher-level examples include measuring the performance worth of a player (e.g., how often the batsman gets on-base) to how well a player is performing compared to the rest of the league. Very few sports organizations have taken that final step and embraced data mining techniques. While the introduction of data mining to the sports environment has been rather recent, the impact of teams using these techniques has been dramatic. In professional baseball, the Oakland Athletics (A's) have long been at the low-end of the league in terms of salaries paid to their players. It was thought that the amount of salary spent on players had a direct correlation to the success of a team, where the more money spent on players would translate into more wins for the organization. In the early 2000s the A's began to embrace data mining as a way of remaining competitive and began a period of success which either put them in the playoffs or in playoff contention year after year despite their low payroll. The Boston Red Sox achieved similar success when they embraced data mining. After going 86 years without a World Championship, the Red Sox won the World Series twice recently (2004 and 2007) and can attribute their success to data mining techniques. The focus of the rest of this monograph will be on the upper levels of relationship between sports and sports data. In particular we will examine a myriad of sport-related statistics as performance measuring devices as well as various data mining techniques.

While the use of statistics in the decision-making process is certainly an improvement versus instinct alone, statistics by themselves can be misleading without an understanding of their fundamental meaning. This misleading tendency of statistics can come from either an imprecise measurement of performance or an over-emphasis of particular statistics by the sports community. As evidence, consider the fact that certain players can build impressive individual statistics yet have little impact on the performance of the team. Sports statistics can suffer from impreciseness where a statistical metric does not truly measure performance contribution. An example of both the impreciseness and over-emphasis by the sporting community with regard to a statistic can be best illustrated by baseball's Runs Batted In (RBI) metric. The RBI has been long been considered the cornerstone of evaluating player contribution. It was created by British-born journalist Henry Chadwick during the mid-1800s, and was an attempt to assign credit to players that directly generate runs (Lewis 2003). Chadwick was more familiar with the game of cricket rather than baseball and applied his knowledge of cricket to the creation of baseball statistics. Even though Chadwick had an incomplete understanding of the nuances in baseball, it was because of his position as a journalist that he was able to popularize his statistics which were never seriously questioned until the latter half of the 20th century. The imprecise nature of the RBI statistic can be summed up in the following thought experiment. Suppose there are two players with the exact same batting average, which means that they both hit the ball and get on-base with the same percentage of success. Let's further imagine that both players routinely hit for singles, advancing themselves and their teammates one base at a time. The RBI statistic for both of our players depends on the actions of those players who batted before them. If team members are able to routinely get on-base for the first of these players and not for the other, then the first of our hypothetical players would be credited with RBIs when their teammates cross home plate as a consequence of the our hypothetical player's hits. The second of our hypothetical players would not receive any RBIs, even though both players performed the exact same actions. Basing a player's worth on the RBI statistic would be a misleading indicator of performance yet both sports organizations and media outlets typically focus on RBI run production as a measure of success. This over-emphasis on the RBI statistic by the sports community makes it difficult to accurately value player performance and make player-based comparisons.

Baseball is not the only sport that is susceptible to problems of statistical imprecision. In basketball, the defensive rebound is the number of times a defensive player acquires the ball from a missed shot attempt. In order to get a defensive rebound, teammates must block out opposing players and by doing so, those players blocking typically cannot get the ball. However, their actions arguably make them just as important in the accomplishment of retrieving the ball . Given the manner in which rebounds are measured, only the player who takes possession of the ball is credited with the rebound.

Besides impreciseness and incorrect use, another difficulty with sports statistics is how to quantify risk. In American football, a defensive back can either wait in mid-field in an attempt to intercept the ball from the opponent or play solid cover defense, where the defensive back seeks to severely limit the opposing team's ability to move closer to their goal. When the defensive back is in mid-field, the player is taking a risk which if successful can quickly change the momentum of the game. If the defensive back plays solid defense, the player is not assuming much risk but rather is playing it safe. However, if the player is successful at taking risks and making interceptions, there is a greater perceived player value. Quantifying risk taking behavior is a difficult problem.

In this monograph, we propose a Sports Data Mining framework which can be used to categorize the different methods that sports organizations use to measure player contribution and uncover new knowledge. From this, we will highlight the problems with existing measurements and how organizations are creating new performance metrics. We will also showcase many of the techniques that organizations are using to make sense of their collected data. By properly leveraging Sports Data Mining techniques, organizations can reap the results of better team performance, identifying contributions, and exploiting the tendencies of opposing players.

For these reasons, it should be no surprise that the sports organizations that have embraced these data-centric techniques are enjoying marked success. The traditional decision-making approach of using intuition or gut instincts is quickly becoming outmoded. Instead, assessments are being made on the basis of strong analysis and scientific exploration. With more and more sports organizations embracing the digital era, it may soon become a battle of the better algorithm or performance metric, where the back-office analysts may be just as important as the players on the field.

2 History

Sabermetrics, credited to Bill James, is often considered one of the earliest attempts to capture baseball performance metrics. Oakland A's General Manager, Billy Beane, fielded a competitive professional baseball team by adopting a sabermetric approach in selecting players in 2002.

The practice of adopting data mining in organized sports was not an overnight phenomenon. Instead, there were many incremental movements that took place over a period of several decades. In baseball, the fundamental shift away from traditional statistics can be credited to Bill James. In 1977, James began publishing his annual *Bill James Baseball Abstracts*. These abstracts were used as his personal forum to question many of the traditional baseball performance metrics and also a place to provide commentary regarding problems with existing imprecise performance measurements. In spite of selling only 50 copies of his abstract initially, James continued to publish his annual compendium of insights, unorthodox ranking formulae and new statistical performance measures which he called *sabermetrics*. Readership started to slowly grow and subscribers to the *Bill James Baseball Abstracts* became interested in the new performance metrics and a discussion began on refining existing metrics. James best described the concept of sabermetrics in the following quote from his 1982 annual of the *Bill James Baseball Abstract* (James 1982).

> Sabermetrics does not begin with the numbers. It begins with issues. The numbers, the statistics are not the subject of the discussion... The subject is baseball. The numbers bear a relationship to that subject and to us which is much like the relationship of tools to a machine and to the mechanic who uses them. The mechanic does not begin with a monkey wrench; basically, he is not even interested in the damn monkey wrench. All that he wants from the monkey wrench is that it do its job and not give him any trouble. He begins with the machine, with the things which he sees and hears there, and from those he forms an idea – a thesis – about what must be happening in that machine. The tools are a way of taking the thing apart so he can see if he was right or if he needs to try something else.

However, sabermetrics was seen as a peripheral endeavor or an academic exercise that was quite foreign to the traditional decision-making formulae of the day. The proposed methods hadn't been thoroughly tested, so because of their novelty and unproven nature, many sports enthusiasts and organizations were resistant to embrace these changes. In order to combat these problems, some sabermetricians began to test sabermetrics by applying them to the operation of fantasy baseball teams. From drafting players for their teams to the day to day managerial decisions, these pioneers experienced overwhelming success versus their traditional peers. Soon other sports enthusiasts and even some sports-media personalities began to embrace sabermetrics as a way to better evaluate player performance and began to vocally promote its adoption. Even with this tidal wave of fan excitement for sabermetrics, sports organizations were still resistant for several decades because the traditional decision-making approach was still deeply entrenched within the business (Lewis 2003).

As the idea of sabermetrics grew and began to cross over to other sports, it wasn't until 2002 that sabermetrics became incorporated into professional baseball. The Oakland A's General Manager, Billy Beane, was searching for ways to field a more competitive team and decided to adopt a sabermetric approach to select players in the draft versus the traditional method of relying on organizational domain experts, the scouts. While the scouts were initially reluctant to their reduction of decision-making capacity, Beane's use of sabermetric tools in the 2002 player draft landed the A's in either the playoffs or playoff contention for 5 straight years (Lewis 2003). Beane discovered that by carefully selecting players in the draft, the A's could lock-in players that were oftentimes overlooked by other clubs, into long contracts that paid little money and thus develop this into a strategy to compete with larger payroll teams. It was simply a matter of picking the right players, which sabermetrics could make easier. While other teams focused on traditional metrics such as RBIs, pitching velocity and subjective measures of whether the player had the body of a professional baseball player, Beane instead looked at measures that were directly applicable to player effectiveness. Measures such as On-Base Percentage (OBP) or how often the batter got on base, and strikeout to walk ratios for pitchers were unconventional and not even considered by other professional teams. As a result, Beane's ranking of desired draft players was completely detached from other clubs. This meant that the players could be acquired much more cheaply and signed into long contracts. As the players progressed and their contracts were about to expire, the A's would then have the option of trading or selling them to the larger market teams and thus receive a return on their investment. This return would then be used to acquire players from other organizations and help keep the team competitive. Relying instead on computers and algorithms to pick talent, the A's produced star players such as Barry Zito, Mark Mulder, Tim Hudson, Jason Giambi, Miguel Tejada, Eric Chavez, Nick Swisher and Mark Teahan.

Up until Beane's entry of sabermetrics in the player draft, the draft was seen as unpredictable because teams never really knew how a drafted player would perform. Clubs usually treated the player draft as a minor event and left the majority of the drafting decisions to their scouting departments. By contrast, Beane systematically analyzed the potential draft picks by the statistics they generated throughout their careers. The first step in Beane's approach was to eliminate all high school players from consideration. This was a significant departure from tradition, where high school players were seen as valuable commodities by other clubs. However, players at the high school level rarely panned out and making comparisons between high school players and leagues at that level was difficult. Instead, Beane focused on college players, as they were more mature, had a better history of developing into major league talent and it was easier to compare performances across the different collegiate leagues. Beane and his colleagues would then use sabermetrics to rank order the draft players. Furthermore, most of the players who were rated highly using these methods were overlooked by other teams. The results soon became clear when Oakland fielded competitive teams year after year in spite of its low payroll. Competitors and commentators did not understand how Oakland was able to win consistently. Even Major League Baseball's Blue Ribbon Panel of economic experts that was investigating the salary inequities in baseball, concluded that Oakland's performance was a statistical anomaly (Levin et al. 2000). At that time, sabermetrics and data mining techniques were not widely accepted in sports.

Baseball was not the only sport that was undergoing a statistical reformation. During the 1980s, Dean Oliver began to ask similar questions about basketball metrics. Oliver, a contemporary of baseball's Bill James, was more interested in creating team-based statistics that centered on team performance rather than individual performance. Like James, Oliver took a similar path and published his thoughts and creative performance measures for the rest of basketball community (Oliver 2005). Through his work, analysts were better able to identify player contribution and even probe into the subjective area of team chemistry by measuring how well certain players perform with one another. Both Oliver and James were eventually recognized by professional sporting organizations as essential sporting analysts and were hired as consultants to the Seattle Supersonics and Boston Red Sox respectively; thus cementing the role of statistical analysis in sports.

Following the performance measurement revolution, data mining quickly entered the scene and was used as an extension of decision-making. Many different sports started to take a serious look at data mining as a source of competitive advantage. While some sports are applying statistics and data mining as an extension to current practices, others have adopted novel and unorthodox approaches. One example is a biomedical injury prediction tool piloted by the Italian professional soccer club, AC Milan. This tool uses software to monitor the quality of player workouts and compares the results against a baseline. Any workout that falls below the baseline expectation, could signal that either an injury has occurred that the player did not disclose or that an existing injury has worsened (Flinders 2002). A second example of novel data mining use is software developed by Las Vegas Sporting Consultants (LVSC) to monitor the betting on sports to look for any unusual wagers that may indicate a corrupt match (Audi and Thompson 2007). The LVSC not only looks for statistical outliers in betting, but they also monitor matches and look for evidence of corrupt officiating and players that engage in point shaving schemes by comparing their performances to prior occasions. A third example of novel data mining research comes from the discovery that physical aptitude correlates to anticipated physical performance (Fieltz and Scott 2003). Every year the National Football League (NFL) conducts a "Combine" where prospective players must engage in a series of physical drills in front of the league's scouts and coaches. Besides the physical challenges that participants must endure, the Combine also assesses the intellectual capacity of prospects through the Wonderlic Personnel Test. The NFL has developed expected Wonderlic scores based on the amount of intelligence required to play a particular position; e.g., a quarterback who has to make a myriad of on-field decisions should have a higher Wonderlic score (24), than a halfback (16) whose job is to run the ball (Zimmerman 1985).

Professional sports are a big business. While revenue from fan attendance has always been a key element, teams that advance to playoffs and/or win championships not only increase attendance but can also draw on additional revenue from lucrative television broadcasts and vast merchandising opportunities. The key is simple: win. With such a dizzying array of competitive forces and an abundance of decisions to be made, it becomes principally important that the right decisions are made in order to maintain a competitive advantage. These decisions come from the hard facts and data already acquired. It is just a matter of finding ways to unlock the knowledge trapped within data and use it.

3 Societal Dimensions

Professional sporting organizations are not the only stakeholders of success. Fans, a nation's citizens, fantasy sporting enthusiasts, and analysts can all have an investment in a team's performance.

The sports fan could be considered one of the most apparent stakeholders of team success. Fans of a sports club can help increase team revenue by attending games, buying franchise merchandise, purchasing ancillary items such as online broadcasts, special access passes and help promote the team to others. Aside from these important contributions to the financial well-being of a team, the sports fan also holds another significant role; the ability to motivate their team and help contribute to winning. This can be clearly seen from the notion of "home field advantage," where the team's fans can exert a certain degree of influence over game outcomes. Keeping the fans in the stands is a tricky proposition. If a team wins, they acquire additional fans. These fans can then help motivate the team to win more, thus delivering more fans to the organization. The secret is to begin the process of winning.

Aside from team-specific fans, sports can also be a source of national pride. Through matches in world competitions such as the FIFA World Cup and the Olympics, national teams can captivate a state's psyche and bring about a feeling of national pride, promote tourism, raise funds to support other teams, sell team-related merchandise and inspire a younger generation to pursue similar goals. One such example can be seen in the sport of soccer. The FIFA World Cup tournament is played every 4 years and arguably has the most passionate sports fans. Fans will wave their nation's flag, sing songs of national interest, paint their faces in their nation's colors and create an atmosphere of national excitement. However, sometimes this excitement can spill over into violence. Fights between rival fans are unfortunately becoming more commonplace. In one unfortunate incident during the 1994 FIFA World Cup match between Colombia and the United States, Colombian player Andrés Escobar inadvertently made a goal for the United States and cost Colombia the match. Escobar, an all-star for the Colombian team, was shot to death soon after in retaliation for his goal in the World Cup match (Almond 1994). However violence is more the exception rather than the norm. Other examples of national pride include unlikely entries and underdogs. The Jamaican bobsled team is a good example of both an unlikely entry and an underdog. Although the warm climate of Jamaica does not appear to be the ideal locale for a winter sport such as bobsled, Jamaica qualified for the 1988 Winter Olympics in Calgary, Alberta. Their team was unexpectedly in medal contention until their sled broke and was carried across the finish line. Their determination earned Jamaica respect as a bobsledding competitor as well as captivated much of the world's attention. They later finished 14th in the 1992 Winter Olympics and won a gold medal in the 2000 World Push Bobsled Competition.

National pride can also be created in times of turmoil and anxiety. During the war years of the United States (1942–1945), baseball players of all levels left professional baseball and joined the military. Some did so even though they were at the peak of their career and poised to set records, such as Cleveland's Bob Feller, Boston's Ted Williams and Detroit's Hank Greenberg. Many arguments still surround these military veterans and what their sporting careers would have been like if they hadn't been interrupted by war. However, the expectation at the time was for military-aged men to go to war. Negative societal consequences could ensue as they did for boxer Jack Dempsey when boxing fans accused him of being a draft dodger during World War I. It turned out that Dempsey did try to enlist but was turned down by the Army. However, it wasn't until after the war had ended that Dempsey was able to clear his name.

Another stakeholder in the success of a sports franchise is the sports fan with fantasy teams. Fantasy sports, also known as rotisserie sports, are owner simulation games that use actual player statistics. Fantasy teams are typically organized into leagues with several other owners to mimic the existing sport. Owners then draft or trade for existing players based upon the owner's expectation of how well that player will perform. Player statistics are then collected for each game and translated into points. The league teams are then rank-ordered based upon their summation of points. The key to winning in fantasy sports is to retain those league players that would maximize points. This became the perfect environment for sabermetric testing. Several owners began to adopt sabermetrics as a method of selecting their team players and were met with overwhelming success versus their non-sabermetric counterparts. It would take many more years before sabermetrics would make the jump into professional sports.

4 The International Landscape

American sports are not the only sports that are embracing statistics and data mining. One sport that is starting to make an impact is the international sport of soccer. Billy Beane, who brought sabermetrics to professional baseball's Oakland A's, is now trying to do the same with soccer through Major League Soccer's San Jose Earthquakes franchise. The problem is that soccer does not have the richness of statistics that baseball has. This does not in any way diminish the game. Statistics in soccer did not develop in the same ways as it did in other professional sports, primarily because of the difficulty of measuring player activity and quantifying game-based events. Beane is trying to change that by introducing a plethora of sabermetric styled statistics to the game. Statistics such as number of touches (how many times a player is involved in a play), shot creation (whether the player shoots the ball themselves or participates in a shoot), ball retention (a measure of offensive turnovers), and balls won per 90 minutes (how effective your defense is to picking up a turnover) (Donegan 2008). However, Beane was not the first to attempt to quantify soccer as a series of mathematical models. Professor Anatoly Zelentsov used cleverly crafted computer programs to not only pick players for Ukraine's Kiev Dynamo but also to analyze every game they played. This strategy led the Dynamo to win the Union of European Football Associations (UEFA) Cup in 1975 and 1986 as well as either win or place in the USSR Championship nearly every year through the 1970s and 1980s. Players selected for the Dynamo were subjected to a series of computer tests measuring nerve, endurance, memory, reaction and coordination. For the nerve test, subjects were given a screen with a vertical line and ten dots of varying velocity. The concept was simple, press a key when the dot traverses the line. It was theorized by Zelentsov that a nervous subject who missed placing one of the dots would overcompensate on the next and hence lower their score. For the endurance test a subject would rapidly press a key to establish a maximum typing speed and then be asked to maintain that maximum speed for an additional 40 s. In the memory test, subjects were given nine numbers under one hundred and had to not only memorize the numbers but their location on the screen. It was believed that subjects that could excel at the memory test would perform better on the field by remembering where other players are going to be during set plays. In the reaction test, the computer screen would quickly flash white pulses of light and the subject was asked to press a key as fast as possible. Finally in the coordination test, subjects were shown a dot traversing a maze and were asked to retrace its path (Kuper 2006). Once players were selected, they were given a series of set plays. The pitch was divided into a series of zones and players were required to be in certain zones during specific plays. This allowed players to pass the ball to others without ever having to see the other player first. Critics of the system likened Dynamo players to soul-less robots that lacked any form of passion. However, their results speak for themselves as the Dynamo was recently ranked number seven of the all-time soccer clubs by the International Federation of Football History and Statistics (IFFHS) (Xinhua News 2009).

Cricket has also started to experience its own foray into sabermetrics. Cricket contains a wealth of data similar to baseball, however, like baseball, the statistics used are not reflective of true performance measures. While data abounds, in such places as the Wisden Almanack which contains Cricket-related match data since 1864 (CricInfo 2008) and CricInfo's Statsguru; the major complaint among sabermetrians is that the data is geared more towards recording data as historical scorecards and not the aggregation of player-specific statistics (Barry 2009; Brewer 2009). To place the data into a sabermetric form requires tedious amounts of manual "cut and paste" or writing a computer program to automatically filter and parse the data. Unfortunately there is no commercially available product to currently fill this need.

John Buchanan, a coach for the Australian National team from 1999 to 2007 and current coach for the Kolkata Knight Riders in the Indian Premier League, understood how to use statistics in cricket for a personal advantage. He noted that many existing statistics in cricket are fundamentally flawed and do not take into account pitch conditions, rule and equipment changes. His personal view is to ignore the existing cricket statistics and instead find those performance measures that provide a better representation of player performance. Secondly, within those metrics find the ones which contribute the most towards winning and match as well as help to evaluate the statistics for any trends of tendencies to exploit (Ball 2008). This has lead to many new sabermetric-style statistics entering into the game of Cricket. Statistics such as Marginal Wins, which started out in baseball parlance, have made the cross-over to cricket. In this statistic an average player's performance is modeled based upon their position and compared against the performance of a specific player. The information gathered can then be used to model ordinary team performance both with or without the player. This information can then be used to help determine whether they are an asset or liability as well as how much of one they are as compared against an average replacement player (CricketAnalysis.com 2009).

Cricket data has also been recently explored using data mining and knowledge management tools to some success. In a study of One Day Test Cricket matches, it was found that a mix of left/right batsmen and a high runs to overs ratio were both highly correlated to winning (Allsopp and Clarke 2004). The usage of alternating left and right-handed batsmen is believed to keep the opposing team's bowler out of their typical rhythm and thus be less effective (Allsopp and Clarke 2004). The high number of runs to overs ratio, (e.g., amount of runs scored as a proportion to the number of offensive periods) indicted that a quicker paced game (i.e., more runs) was also a factor in determining a winning team. These factors can further be used to determine team effectiveness in tournament play.

Hockey is another sport in the infancy of sabermetric research. Several well-known baseball sabermetricians have begun to investigate whether sabermetric techniques can be applied to other sports. Phil Birnbaum is one such example where his insights have begun to be observed in cricket and hockey. However, in the sport of hockey, statistical research is much like it was several decades ago. Although some statisticians attempted to make inroads into the sport, such as Daryl Shilling's Hockey Project (Shilling 2005) and Mike Dunshee's sabermetric hockey system (Dunshee 2007), not much has been done since. In the case of the Hockey Project, Daryl essentially abandoned it. The current state of statistical analysis in hockey is best explained as "utterly maddening" and is summed up in the following argument.

> Think about the key statistic in baseball: the batting average. Did you ever hear a ballplayer being judged by the absolute number of hits he gets (unless it's a lot)? Yet in hockey, a player who plays 7 min a game is routinely described by his "15 goal season" or "he only has six goals in the first half." Some supposed superstars are on the ice 30 min a game. Shouldn't we be looking at "points per minute played"? (Goodman 2005).

5 Criticisms

> Not all sports are currently equal in their application of sports data mining techniques. Even in baseball, sabermetrics has its share of detractors and vocal critics.

Not all sports are currently equal in their application of sports data mining techniques. Some sports such as baseball are clearly farther ahead than others in their treatment of data. The sports that attract high-dollar talent are generally also the ones to be more open to the idea of letting statistics and computer models make managerial decisions. However, even in baseball, sabermetrics has its share of detractors and vocal critics. The most notable critic is baseball hall of famer, turned commentator, Joe Morgan. Morgan believes that player performance is more qualitative where heart, passion and other immeasurable and intangible factors all contribute. Whereas sabermetrics is a quantitative approach favoring hard evidence of measurable performance indicators. Baseball has played under the qualitative approach for over 100 years and in many organizations, still does. However, consider the records of those general managers that instead chose to follow quantitative analysis:

Billy Beane, general manager of the Oakland A's from 1998 to present has guided his team to playoff appearances in 2000, 2001, 2002, 2003 and 2006. When the team didn't make the playoffs in 2004 and 2005, they still had more wins than loses for the season.

JP Riccardi, general manager of the Toronto Blue Jays from 2002 to present, used to work under Beane at the A's. Riccardi hasn't made the playoffs yet in the tumultuous American League East division, however, he guided them to winning seasons in 2003, 2006, 2007 and 2008 placing them in contention against the likes of the Yankees and Red Sox.

Paul DePodesta, former general manager of the Los Angeles Dodgers for the 2004 and 2005 seasons, also used to work under Beane at the A's. While at the Dodgers, DePodesta guided them to the playoffs in 2004 before a series of injuries hurt their chances in 2005. While at the Dodgers, DePodesta made several quantitative trades which irked fans and the media and it is suspected that incessant sabermetric-opposition in the local media led to DePodesta's departure.

Theo Epstein, general manager of the Boston Red Sox from the 2003 season to present, was also an acolyte of Bill James' work. He guided the Red Sox to five playoff appearances in a 6 year span, winning the World Series twice. In 2006, although the Red Sox did not make the playoffs, they still had a winning record.

While sabermetrics may have its critics, it is hard to argue against the success enjoyed by those that understand and use it, placing those teams in the playoffs or in playoff contention year after year. However, this issue has passionate supporters and detractors. As a result, it may take many more years of hard data to finally settle the case.

6 Questions for Discussion

1. Is the sabermetric approach the correct approach? Are there other performance metrics that you feel are equally as important?
2. Are there any sports that would not benefit from a data mining approach? Why?
3. How can researchers introduce academic findings to professional organizations or fans and demonstrate their value?

Chapter 2
Sports Data Mining Methodology

Chapter Overview

Data Mining involves procedures for uncovering hidden trends and developing new data and information from data sources. These sources can include well-structured and defined databases, such as statistical compilations, or unstructured data in the form of multimedia sources such as video broadcasts and play-by-play narration.

Data mining activities, the tools, technologies and human expertise, are rooted within the field of Knowledge Management (Davenport and Prusak 1998). Knowledge Management can provide an organization with a means of competitive advantage (Lahti and Beyerlein 2000) and a method for maintaining the continuity of knowledge in the organization (Serenko and Bontis 2004). Through knowledge sharing and retention, businesses are discovering increased productivity and innovation (O'Reilly and Knight 2007). However, before getting raw data to a stage of useable knowledge, we must first examine the intermediate levels of data and knowledge as represented by the Data-Information-Knowledge-Wisdom (DIKW) hierarchy (Ackoff 1989). The DIKW hierarchy is a widely accepted concept in knowledge management circles as a way to represent different levels of what individuals see and know (Cleveland 1982; Zeleny 1987). Each successive level; data, information, knowledge and wisdom, builds upon prior levels and provides an increased awareness of surroundings (Carlisle 2006) where meaning can be found (Chen 2001, 2006).

The DIKW framework then sets the stage for disambiguating data from knowledge and sets definitional boundaries for what data, information and knowledge are. Applying this to the sports domain, certain activities and techniques operate at the data level (i.e., data collection, data mining and basic statistics). Other techniques and algorithms are more suited to the knowledge end of the spectrum, such as strategies and simulations. Throughout this chapter, the DIKW framework can be used to identify the set of relevant tools that can be used depending whether data or knowledge is desired.

R.P. Schumaker et al., *Sports Data Mining*, Integrated Series in Information Systems 26, DOI 10.1007/978-1-4419-6730-5_2, © Springer Science+Business Media, LLC 2010

1 Scientific Foundation

Data mining research can trace back to three distinct scientific disciplines: statistics, artificial intelligence, and machine learning.

While the term "data mining" did not gain widespread acceptance until the 1980s, and has been around in various forms. Its ancestry can be traced back through three distinct disciplines and are still very apparent in data mining research today: statistics, artificial intelligence, and machine learning (Chen and Chau 2004).

In statistical research, data mining evolved as a method to find the reasons behind relations. From statistics, we can find and measure the strength of a relationship (e.g., co-variance) between two variables from the data. But this statistical measurement by itself is unable to explain why the relationship exists or what possible impact it may have in the future (DataSoftSystems 2009). Data mining provides us with the tools to interrogate the data and gain further knowledge about the dependency relationships. It does so through an interactive, iterative and/or exploratory analysis of the data. From the statistical branch of the data mining genealogy, methods such as regression analysis, discriminant analysis and clustering became data mining tools (Data Mining Software 2009).

In regression analysis, data points are examined by fitting a line or polynomial to explain the most data while at the same time minimizing the fitting error. From a regression analysis, causal relationships can sometimes be discovered between dependent and independent variables. Most oftentimes, this type of analysis is used for prediction where if the observed trend holds true, estimates of the dependent variable's value can be made by varying the values of independent variables.

Discriminant analysis is a classifier-based approach which seeks to best categorize the data based upon the combination of features that contribute to maximally separating the data. This analysis can not only identify the important classification features, but also provide insights into the relationship.

Clustering is another classifier method which seeks to partition the data into different sets based on their features. Clustering can take many forms including hierarchical clustering, or sub-partitioning of data, density-based clustering where item similarity can be measured, and clustering based on Euclidian distances where the distance between two clusters can provide a measure of similarity between them.

The second branch of data mining, artificial intelligence differs from statistics by applying a heuristic algorithm to the data. Heuristic-based approaches are experience-based techniques that can be computationally intense. It wasn't until the 1980s that computers began to possess sufficient power to handle the complexities associated with heuristics. This approach attempts to balance out the statistics by applying a human problem-oriented perspective to problem solving based upon educated guesses or common sense. Imparting this problem-solving technique to a computer is a matter of learning from appropriate rules or cases. Heuristic solutions may not be perfect, however, the solutions generated are considered adequate.

Rule-based heuristics are a set of conditional statements that the computer tests in sequence in order to obtain a solution. This style of problem solving became popular during the 1980s in expert systems. Rule-based heuristics can derive new knowledge from the existing problem through inductive reasoning.

Case-based heuristics, or case-based reasoning, approaches problems based on solutions to past problems. This style of decision-making relies on having a sufficient breadth of historical cases and solutions to draw upon in order to create solutions by analogy.

The third branch, machine learning, uses algorithms to learn and induce knowledge from the data (DataSoftSystems 2009; Chen and Chau 2004). Examples of these algorithms include both supervised and unsupervised learning techniques such as genetic algorithms, neural networks, self-organizing maps and Bayesian methods. These techniques can iteratively identify previously unknown patterns within the data and add to our understanding of the composition of the dataset.

Genetic algorithms (GAs) are a stochastic optimal search technique based on evolutionary processes, where parallel solutions can evolve through crossover and mutation. In a GA, a number of initial solutions are created, tasked with the overall goal of finding an optimal solution. When one generation has finished, the solutions are then allowed to pair up and create offspring solutions for the subsequent generation. Not all solutions will participate in this procreation process as it is governed by chance. However, stronger solutions are given more chances than weaker ones to make it into the next generation. The solutions that have been selected then perform a cross-over to build offspring solutions and the parents are then removed. Also, as with evolutionary development, a small number of mutations are permitted which can sometimes have the effect of creating a much better solution by accident. This indeterminate process is then allowed to continue until a stopping condition is met which could be a preset number of generations, or more commonly the point where further generations do not appear to produce better results.

Neural networks are another machine learning technique that can learn to classify from data patterns. The most common form of Neural Network is the Back Propagation Neural Network (BPNN), which is a supervised learning technique that adjusts itself to the data patterns and can be used to make predictions from unseen data. BPNNs attempt to simulate human synaptic activity by modeling the brain's mesh-like network of neurons and axons as perceptrons and weighted links. From adjusting the weighted links based on the training classes, BPNNs can recognize complex data patterns and learn to anticipate them.

Self-organizing maps (SOMs) are a special type of neural network that are used to visualize multidimensional data in lower dimensions, typically a two-dimensional plane. This visualization can help to identify trends or tendencies that can be clouded by higher orders of dimensionality.

Bayesian statistics considers the probability of states given the data. These probabilities include prior probability, a measure of the probability of a state condition being met before the introduction of data, conditional probability of seeing that data if the state is correct, the marginal probability of the data considering all states and posterior probability that measures the belief in the state condition being met after seeing the data. This technique can then provide a measurable value of possible states given the data.

2 Traditional Data Mining Applications

Both structured and unstructured data can be "mined" for knowledge in different scientific, engineering, business, biomedical, public safety, and sports applications.

Data mining in a business environment is a common way of obtaining organizational intelligence. Traditionally, data mining within business has been concerned with capturing new knowledge within the data. This data could be somewhat structured yet raw, such as retail purchases, or be unstructured and difficult to computationally parse, as can be the case with textual data. Whether it be structured or unstructured, once the data has been parsed, cleaned and organized, data mining can take place.

Examples of data mining discovery on structured data routinely include using the customary grocery/retail illustrations. Beer and diaper sales have long shown a stronger than expected correlation in sales. While statistics by themselves can be used to identify these types of relationships, such as covariance or ANOVA-type analyses, the statistics cannot explain why the relationship exists, simply that it does. This is where data mining becomes an invaluable tool, by allowing us the answer the why type of questions and in the process gain new insights into the relationship. As it turns out, data mining the customer demographics of these sales discovered that men sent to the store (by the wives) to buy diapers, would also pick up a case of beer. Once discovered, some grocery stores changed their product layouts such that beer and diapers were located proximate to one another, in some cases within the same aisle, in order to increase impulse consumer purchases. A second example of effective data mining usage in the retail sector comes from Walmart's ability to track colds and flu more quickly and precisely than the US government's Center for Disease Control (CDC). Walmart employs a sophisticated product inventory and reordering system that is primarily data driven. Using this product data, Walmart knows what the baseline sales of cold and flu remedies should be for their stores. When sales of these products exhibit a statistical spike in sales, a cold or flu outbreak has occurred within that store's sphere of influence. Congregating the data together and adding in the geographic locations of their stores, Walmart can not only see the spread of colds and flu from store to store, but also can predict future affected stores and therefore stock their shelves appropriately. As a third example of structured data mining usage, and an example outside of the retail arena, police reports are a good example of well structured documents that can be used to detect criminal aliases. When criminals do not want to be positively identified, they will provide false information such as name, address, date of birth, etc. In order to combat the introduction of false information, police will routinely ask for the information multiple times. A careless criminal may not remember their previous lie and get caught by this tactic. Criminals that know this tactic, will provide information that is close to the real data so that they can remember their falsehood and perpetrate it successfully. This misdirection can lead to multiple criminal aliases which can be laborious for an officer to detect. By introducing textual tools that comb through police reports and look for data that is close, but slightly off, the officer's workload is reduced and criminal deception can be discovered (Chen 2006).

Unstructured data, and notably textual data, is more difficult to manage. Processes must be put in place to properly extract the data from these sources. To do this, there are two major schools of thought; a template-based approach where data is identified and pigeon-holed into particular fields using a template, or a lexical/syntactic or semantic analysis of the textual structure or meaning of the text. Template-based approaches work great when there is some form of order to the text. Sales receipts, product invoices and shipping documents all fit this definition because their defined structure allows a textual parser to effectively extract the data. Memos, letters and newspapers articles do not conform to template-based methods and require a different type of analysis. In a lexical/syntactic analysis, the structure of the text is examined. Terms are categorized by their parts of speech using either a lexicon or dictionary, or identified by the syntax of the sentence. Examples of using this type of analysis include predicting stock prices from financial news articles, identifying specific cancer gene pathway associations in medical abstracts and identifying unknown authors by their writing style tendencies (Chen 2006).

In predicting stock prices from financial news articles, terms within news articles give us clues as to the probable price movement of a company. Terms such as "factory exploded" or "workers strike" typically indicate a downward price movement while "unexpected gains" and "earnings surprise" may indicate upward trends. By analyzing the article terms against price movements, stock prices can be predicted with better than chance accuracy immediately following the release of the news article. Another example includes identifying relations in cancer gene pathways from medical abstracts. Oftentimes, medical researchers will focus their attention on particular subsets of cancer research. Whether it be an enzyme reaction, or observed effects of different genes, researchers write about their experiment and move on. The problem is, with so many researchers focusing on the minutia of cancer, there isn't a lot of work on the bigger picture, or putting the dots together and getting a more complete picture of the disease. However, it is quite difficult to achieve this, because it would require a medical researcher to become familiar with a majority of current research as well as possess an ability to put it all together. This problem is well suited to textual data mining and machine learning techniques that can identify the gene pathways described in medical abstracts and find previously undiscovered links. A third example includes discovering the identity of authors based upon previous writing samples. All individuals have certain tendencies within their writing styles. Whether it be how a sentence is constructed, particular word choices or punctuation usage, these indicators can be used to identify the similarities between an unknown author and a body of known ones (Chen 2001, 2006).

A semantic analysis instead takes aim at the meaning of the terms by applying either a hierarchical framework, as is the case with WordNet and Cyc, or looks at the context of sentence for clues to problems such as word-sense disambiguation. Applications that employ Cyc, a type of hierarchical encyclopedia for computers, can gain additional insight not otherwise obtained. As an example of using this technology, an application that came across the phrase "lemon tree" could query Cyc and traverse the hierarchy upwards to discover that lemon tree is a type of citrus tree which is a type of deciduous tree which is a type of tree which is a type of plant. The application could also seek a more detailed explanation by traversing the hierarchy downwards and discovering that Lisbon lemon is a specific instance of a lemon tree. In word-sense disambiguation, some terms may be ambiguous to the computer. Bank is one such example, where the term could mean a financial institution or an adjoining land area proximate to a waterway. By analyzing the terms surrounding the ambiguous term, clues can be discovered to distinguish its meaning.

3 Deriving Knowledge

At each level of the Data-Information-Knowledge-Wisdom hierarchy, meaning can be obtained through filtering and analysis, and data can tell a story and explain the observed phenomenon.

Returning to the DIKW hierarchy, each level can provide specific meaning to describe the dataset. This meaning is obtained through filtering of relevant material and abstracting it in such a way that the data begins to tell a story and explain the observed phenomenon.

Data are the observable differences in physical states (Boisot and Canals 2004) that are acquired from stimuli and examination of the world around us. By themselves, data are generally overwhelming and not entirely useable. In our framework, data can be thought of as all of the individual events that took place in the sports match. If applied to baseball, this data could contain a record of pitch sequences, at-bats and defensive moves which by themselves provide little interest or value.

To be of practical value, data needs to be transformed by identifying relationships (Barlas et al. 2005) or limited to only that which is relevant to the problem at-hand (Carlisle 2006). This data transformation produces information which can be characterized as meaningful, useful data (Bierly et al. 2000). Returning to our baseball example, information could focus on the pitch sequences by a particular pitcher or batting sequence of a certain batter. Although information is not very useful at this stage, abstracting it to the next level of the hierarchy, knowledge, can provide additional meaning by identifying patterns or rules within the data.

Knowledge is the aggregation of related information (Barlas et al. 2005), that forms a set of expectations or rules (Boisot and Canals 2004) and provides a clearer understanding of the aggregated information (Bierly et al. 2000). At this level of the DIKW hierarchy, rule-based systems can be developed to aid individuals in both expanding their own knowledge while also providing a benefit to the organization (Alavi and Leidner 2001). Using our baseball example again, analysts can evaluate the pitching information of a particular player and look for tendencies or expectations in the types of pitches thrown. Data mining can then be defined as the pursuit of knowledge within the data.

Precise definitions of data, information and knowledge are still a matter of debate within the Knowledge Management community. So is it with the final level of the DIKW hierarchy, wisdom. Wisdom can be viewed as a grasp of the overall situation (Barlas et al. 2005), that uses knowledge and knowledge alone (Carlisle 2006) to achieve goals (Bierly et al. 2000). In our baseball example, we have several disparate pieces of knowledge as well as an ultimate goal. From data mining we can obtain knowledge of the types of pitches to be expected, knowledge of effective strategies to approach specific types of pitches and an overriding goal that a successful at-bat can help win a game. Putting all of this knowledge together into wisdom; the batter has a chance to positively influence the game in their favor if the observed pitching data pattern holds and the batter is able to use it to his advantage. Wisdom resides in the capabilities of cognition and human understanding (Carlisle 2006), as a computational approach is currently difficult to capture (Barlas et al. 2005).

4 Questions for Discussion

1. Can you see any circumstance where data mining of sports data may not be beneficial, and if so, is it because of problems with performance metrics or data mining itself?
2. Choose a sport and analyze how the three branches of data mining (statistics, artificial intelligence, and machine learning) could be used to derive knowledge.
3. Is there any particular sport which lends itself better to one particular branch of data mining, and why?

Chapter 3
Data Sources for Sports

Chapter Overview

Data, the life-blood of modern sport analysis, has undergone its own revolution. It used to be that data was simply viewed as a record of the game's events that was kept either by the organizations or the responsible leagues for historical purposes. That data became transformed into a condensed form to provide a brief recap of the game's events through a newspaper boxscore. It wasn't until many years later that publishing of data became cheap enough to fill a growing niche of interest. Game data was then expanded upon with comparisons made across different sets. This activity led to refinement as new ideas were introduced of what data should be captured. Then with the advent of the Internet revolution, data rose to the height of accessibility, where sport-related data could be found easily and quickly, oftentimes in searchable form.

1 Introduction

Sporting organizations, professional societies, sport-related associations, and special interest sources have collected rich data resources that can be used for sports data mining purposes.

Sport-related data can come from a variety of sources. The most typical of which is a statistician employed by a sporting organization to record both team-level and individual player performances. Because many organizations keep the data for themselves, third-party professional societies and application-specific companies have filled the gap by providing data sources for sporting enthusiasts and sometimes to the sporting organizations themselves. This in turn has led to

R.P. Schumaker et al., *Sports Data Mining*, Integrated Series in Information Systems 26, DOI 10.1007/978-1-4419-6730-5_3, © Springer Science+Business Media, LLC 2010

the development of sports derivatives such as performance tracking systems, fantasy sport leagues and more realistic gaming based on actual sports data. Professional societies, sport-related associations and special interest sources have filled many of the sports-related data gaps and provided a richness not otherwise available.

2 Professional Societies

A number of professional societies offer sport-related data and act as a community forum to share and explore their knowledge. They mostly serve as centralized repositories where members can share insights and conduct further research. Many of these societies will collect, evaluate, store and disseminate sport-related data for members as well as maintain periodical newsletters and journals. However, their main activity revolves around discovering and sharing knowledge within the sporting community.

2.1 The Society for American Baseball Research

The Society for American Baseball Research (SABR) was formed in Baseball's Hall of Fame Library in August of 1971 (Society for American Baseball Research 2008). Its mission is to foster research about baseball and create a repository of baseball knowledge not captured by the box scores, all while generating interest in the game. While most SABR research concerns itself with insights into particular players or compiled histories of leagues (e.g., the Negro Leagues), a minority of research is quantitative and deals with number crunching of performance data. This line of research has come to be known as sabermetrics and started in 1974 when SABR founded the Statistical Analysis Committee (SAC). This committee is charged with the goal of carefully studying both the historical and modern game of baseball from an analytical point of view. The SAC Committee publishes its research on a quarterly basis and presents their key findings at annual SABR conventions (Birnbaum 2008).

2.2 Association for Professional Basketball Research

For basketball, the Association for Professional Basketball Research (APBR) was formed in 1997 with the objective of promoting the history of basketball as well as analyzing the statistics of the sport objectively (Solieman 2006). While APBR research mainly concentrates on NBA-related statistics, they also include data from rival basketball leagues, many of which are now defunct (The Association for Professional Basketball Research 2008). Similar to Baseball's Sabermetrics, the APBR has developed APBRmetrics which are used to create better measurements and statistical yardsticks for comparison purposes. APBRmetrics was actually born from much of the early work done in baseball's sabermetrics. Early work by the APBR concentrated on Linear Weights type of metrics where weights are assigned to critical statistics as a way of measuring performance. However, during the 1990s the APBR and Dean Oliver in particular, began the investigation of possession and team-related statistics which has since defined the APBR as the premier source for quantitative basketball research.

2.3 Professional Football Researchers Association

The Professional Football Researchers Association (PFRA) was started in 1979 with the goal of preserving and reconstructing historical game day events (Professional Football Researchers Association 2008). The PFRA publishes articles on a bi-monthly basis, which cover statistical analyses as well as new methods of performance measurement. While lacking an official committee devoted to statistics and measures of performance, members are able to share their resources and insights within the organization.

3 Sport-Related Associations

In addition to professional sport-related societies, associations also collect and disseminate information to their membership. These associations differ from professional societies in that they are not aligned with any particular sport, but with a particular goal in mind, such as improving on existing techniques or archiving collected material for future generations.

3.1 The International Association on Computer Science in Sport

The International Association on Computer Science in Sport (IACSS) was founded in 1997 to improve the cooperation amongst international researchers interested in applying Computer Science techniques and technologies to sport-related challenges (International Association on Computer Science in Sport 2008). The IACSS focuses on disseminating the research of their members through periodic newsletters, journals and biannual conferences.

3.2 The International Association for Sports Information

The International Association for Sports Information (IASI) was founded in 1960 with the goal of standardizing and archiving the world's sports libraries (International Association for Sports Information 2008). The IASI is a worldwide network of sport experts, librarians and document repositories. The Association's information dissemination comes in the form of a tri-annual newsletter and an organized World Congress every 4 years.

4 Special Interest Sources

In addition to sport-related societies and associations, there are other, sometimes commercial, organizations that collect and analyze sport specific statistics. Oftentimes these sources offer traditional statistics as well as augmented data in the form of player biographies, records and awards.

4.1 Baseball

Baseball has a rich history of statistical data and third-party sources. Examples of these sources include Baseball-Reference.com which portrays itself as a one-stop shop for all basic statistics, current standings, player and team rankings by various categories, draft picks, and historical box score data (Baseball-Reference.com 2008).

Retrosheet is another baseball resource that focuses on computerizing all professional baseball records for games played before 1984. Users can browse play-by-play text, box scores and even regular season schedules.

4.2 Basketball

Synergy Sports Technology has redefined Basketball accessible data by indexing live video broadcasts into searchable media. Synergy's product *Synergy Online* allows users to query every play of the game, with constantly updating player statistics and direct broadcasts to desktops or mobile devices. Synergy is also driving the data behind gaming solutions such as NBA Live 09 and NBA Live 10. Their signature product, Digital DNA, models nearly 1,000 characteristics and tendencies of a player (Colston 2009) in order to better model their expected performance and bring a realism to the gaming experience (Arnovitz 2009).

This technology can also be harnessed by NBA coaches to identify the most favorable matchups and substitution routines.

82games.com is a similar archive for Basketball-related data that relies on traditional game data statistics. This resource positions itself as Basketball's innovative data source for fans, coaches and the media (82games.com 2008).

4.3 Football

Pro-Football-Reference.com compiles player, team and league stats along with historical game data (Pro-Football-Reference.com 2008). With this reference, users can query data from historical players as well as teams and sort them according to predefined criteria.

4.4 Cricket

While North American sports are steeped in statistical analysis and sabermetric-style analysis, many sports outside of North America are oftentimes overlooked. The sport of Cricket is one such example that has a rich history of data, but not much analysis. One site that is trying to introduce sabermetic analysis to Cricket is Cricket Analysis.com. This site attempts to breakdown existing statistics and introduce new ones as ways of better modeling player and team performance (Cricket Analysis 2009). The models used are still in their infancy and variables used are constantly changing as their predictive techniques improve.

Another Cricket source is the Wisden Almanack which touts itself as the authoritative source on Cricket match data. This compilation features player data, match scorecards and features match data dating back to 1864 (CricInfo 2008).

4.5 Soccer

Soccer is similarly experiencing its own sabermetric revolution with products such as Match Analysis which provides users with a searchable video collection and statistical analysis tools of player and team data. Designed for coaches, fans and the media, Match Analysis has also become the premier data source for soccer clubs, boasting that their members have won every MLS Cup to date as well as 16 national championships in the past 6 years (Match Analysis 2009).

4.6 *Multiple Sports*

AccuScore is a sports forecasting service that is prominently displayed in USA Today newspapers. Accuscore extends itself into a wide variety of sports including Baseball, Basketball, Football and Hockey, both at the professional and collegiate levels. Users of the online version can make comparisons between players, teams, compute likely point spreads, over/unders and use basic betting calculators.

5 Conclusions

The major North American sports have started to see an acceptance of the new performance metrics and data mining tools. Non-North American sports are just beginning to embrace the sabermetrics concept and the data mining approach.

With the introduction of sabermetric-style data analysis several decades ago, an explosion of content using data-driven techniques and tools has reached into almost every sport. The major North American sports have started to see a stability in the metrics used and a general acceptance of the tools as a necessary component for coaches and an added benefit for fans. It is common for organizations and sports fans to use many of the sabermetric resources available to them as a part of the culture. Non-North American sports are just beginning to embrace the sabermetrics concept, as seen in some of the sport-related resources available. As their understanding of performance modeling begins to mature, a similar acceptance will likely take place.

6 Questions for Discussion

1. What are some other sport-related data sources that you may have used?
2. Is there anything you would improve in existing data sources? Why or why not?
3. What problems could arise if data was extracted from multiple sources?

Chapter 4
Research in Sports Statistics

Chapter Overview

This chapter investigates the role that statistics plays in knowledge creation. While many of these techniques have stood the test of time, some have undergone intense scrutiny while others have experienced transformative processes. All the while we must ask ourselves, are we really measuring what we think we are measuring? This chapter will help to make that distinction.

1 Introduction

Once sport-related data has been gathered, the next steps involve a process of finding the knowledge within. Many different types of statistical analyses can be applied to data rich sports such as baseball and basketball as well as less data-intense sports such as Curling. While the techniques and measurements will change from sport to sport, the heart of the matter, the statistics, are identifiable across sports, even if their measures cannot be directly compared. The methodology behind the numbers remains much the same across sports. Some analyses can be used to measure player performance, team balance, opposition weaknesses and even the possibility of a debilitating injury.

2 Sports Statistics

A myriad of statistics have been kept as records of sports events for the greater part of the last century. The statistics themselves were taken for granted and rarely questioned. However, some in the sports world began to ask, are we measuring what we think we are measuring. Early pioneers of statistical analysis such as Bill

R.P. Schumaker et al., *Sports Data Mining*, Integrated Series in Information Systems 26, DOI 10.1007/978-1-4419-6730-5_4, © Springer Science+Business Media, LLC 2010

James and Dean Oliver not only asked these questions, but also began to offer new statistics and insights.

2.1 History and Inherent Problems of Statistics in Sports

Early baseball statistics are often credited to Henry Chadwick, a nineteenth century sportswriter and statistician (Lewis 2003). Many of today's familiar statistics, e.g., batting average and earned run average, owe their existence to him. However, Chadwick had an incomplete understanding of baseball and based many of the baseball statistics on his experience with the game of cricket. This is one of the reasons why walks (i.e., advancing to a base without a hit) are not included in these formulae, because the walk had no equivalency in cricket.

Another statistic that does not take walks into account is batting average, which is defined as the number of hits a player collects divided by the number of times at-bat. If a player receives a walk during their time at bat, then the at-bat does not count. This leads to imprecision when rating players, because if the goal is to get runs and runs are typically made by someone who gets on-base, both hits and walks should be counted. Players who walk often may have lower batting averages. Therefore using batting average as a sole measurement of performance may lead to unfair comparisons and underestimate player contribution to team performance.

Similarly, the Earned Run Average (ERA) is another cornerstone of baseball's performance metrics. The ERA is the number of earned runs against a pitcher per nine innings. The term "earned run" is important because it is a run that is achieved through a hit. Other means of getting on base and scoring, such as getting hit by a pitch (when the batter is awarded first base after being hit by the ball during an at bat), a balk (an illegal motion by the pitcher which results in base runners being awarded the next base), a dropped third strike (normally a batter strikes out after a third strike but can attempt to run to first base if the catcher drops the pitch), fielding errors and walks, do not count towards the ERA. An over-emphasis on hitting tends to skew ERA values and not portray the full story. These two statistics alone, batting average and ERA, were used as the primary performance indicators by scouts, coaches and general managers for well over a century.

Baseball is not the only sport that suffers from performance measurement imprecision. American football also has some measurement imprecision in statistics such as the number of receptions and yards per carry. Keep in mind that the ultimate goal in football is to score points and as an offensive player, the primary way to do that is to score touchdowns. Every statistic should then be analyzed through this prism. However, the number of receptions is the number of times a player catches a forward pass and does not indicate their frequency of success. Receptions may indicate a preference for a particular player by a quarterback and thus inflate the reception total of the preferred receiver. Yards per carry is another example, where success is not predicated on scoring points. Should one player who ran for 40 yards in one play be valued more than another that runs an average of 3 yards per play? While one obvious solution would be to only compare those players with a minimum number of carries, the process of setting arbitrary thresholds ignores the issue that yards per carry does not take into account the points scored, and thus the statistic leads to inexact comparisons.

Basketball also has its share of imprecise statistics such as field goal percentage and rebounds. The field goal percentage is the number of field goals made divided by the number attempted. A player who scored a high number of points while at the same time has a low field goal percentage might be rated as unsuccessful. Likewise, the rebound statistic, or the number of times a player gets the ball after a missed shot attempt, does not imply that points will be scored. Nevertheless, basketball experts have used these statistics as measures of performance.

The problem with traditional formulae lies in what the statistic is intended to measure. Oftentimes, data are gathered and used in ways that cannot be meaningfully interpreted. The data itself is not at fault, it is the methods that are used for comparing player performances. This also leads us to the realization that there are some problems that cannot be answered through statistical examination alone. The questioning of statistics that were held as truths, the very foundations of modern sports, brought about new techniques and measures which have rapidly become commonplace within modern sports organizations.

2.2 *Bill James*

> Sabermetrics, a baseball performance metric, was credited to Bill James. In 2002, Major League Baseball's Oakland A's General Manager Billy Beane became the first GM to adopt sabermetrics in selecting draft picks and fielding a competitive team.

The shift from traditional statistics to knowledge management was the result of persistent questioning and examination of performance criteria by Bill James. James published a series of "Bill James Baseball Abstracts" in which he began to openly question traditional statistics and offer his own unique insight about remedying the problems he was encountering. Readers of the Bill James Baseball Abstracts became interested in the new way of computing performance and began to make their own contributions. Soon sporting enthusiasts and fantasy baseball team owners began to embrace this data-driven approach and apply it with overwhelming success. Even with a tidal wave of fan excitement for this revolution in thinking, sports organizations were quite resistant to these new ideas for several decades because scouting was so entrenched within organization as the sole vessel of knowledge (Lewis 2003).

In 2002, Major League Baseball's Oakland A's General Manager Billy Beane became the first GM to adopt sabermetrics in selecting draft picks. Beane's use of data mining and knowledge extraction tools landed the A's in either the playoffs or playoff contention for 5 straight years (Lewis 2003).

That same year, the Boston Red Sox hired another Bill James disciple, Theo Epstein. Epstein, a Yale graduate, similarly appreciated the hard facts that could be gleaned from reams of data. He hired Bill James as a consultant in 2003 and went on to engineer the Red Sox World Championships in 2004 and 2007.

2.3 Dean Oliver

Outside of baseball, similar questions began to be asked. Dean Oliver, the Bill James of basketball, was one such pioneer. Asking similar questions throughout the 1980s, Oliver sought to better quantify player contribution and began popularizing APBRmetrics, basketball's answer to sabermetrics. He focused his attention on the proper usage of the possession statistic, where possession is defined as the period of time one team has the ball. Part of Oliver's contribution was to evaluate team performance on how many points they scored or allowed opponents to score per 100 possessions. In 2004, Dean Oliver was hired as a consultant to the Seattle Supersonics, ushering basketball into the Moneyball era. Seattle then went on to win the Division title in 2005.

Also in 2005, the Houston Rockets hired Daryl Morey as assistant general manager. Morey, an MIT graduate and believer in knowledge management principles, had previously worked with the Boston Celtics and STATS Inc where he invented and refined several new basketball statistics (MIT Sloan Alumni Profile 2008).

3 Baseball Research

Baseball has been called America's "National Pastime" and has been a part of the culture for close to two centuries. The first professional baseball team, the Cincinnati Red Stockings, was founded in 1869 and played amateur teams across the country, amassing an impressive 81 game winning streak (Voigt 1969). In 1876 baseball then was organized into sustainable leagues with the creation of the National League, which is still with us today. The National League, which quickly became the premier baseball league, withstood competition from rival leagues such as the Player's League, the American Association, and the Federal League (Fetter 2003). However, in 1901 a startup league named the American League was not so easy to vanquish. Both leagues competed intensely with one another and poached talent to the detriment of the game. In 1903, both leagues agreed to recognize each other as major league and began the tradition of World Series competitions between the two rival leagues (The New York Times Staff 2004).

3.1 Building Blocks

Statistics by themselves should not be the primary means by which player performance is determined, but rather the beginning of a process in which useful knowledge can be discovered. For instance, one of baseball's fundamental statistics has been hits. As discussed earlier, this statistic does not account for other means that a player has to get on-base and these additional means do not count in either the player's batting average or hits total. Because of this, On-Base Percentage (OBP) was developed to better measure the player's ability to get on base by including these different methods. Furthermore, another statistic that can better measure offensive player productivity is slugging percentage. With slugging percentage, the number of bases reached is divided by the number of at-bats, and rewards players who hit doubles and triples instead of singles, or hit home runs. By contrast the hits statistic treats doubles, triples and homeruns as equivalent to singles.

Building upon both of the fundamental statistics of OBP and Slugging percentage, we can derive the On-Base Plus Slugging (OPS) statistic, which is the summation of these two statistics and provides a better representation of a player's ability to get on base and hit with power. OPS is considered to be one of the most effective measures of a player's offensive capability.

3.2 Runs Created

In Bill James' third *Baseball Abstract*, James reasoned that player performance should be measured based upon what they are trying to accomplish, scoring runs, instead of using baseball's predominant indicator of the day, batting average (James 1979). James recognized the disconnect between the two concepts and questioned how run production could be better measured. From this, James developed the Runs Created (RC) formula (James 1982).

$$\text{Runs Created} = ((\text{Hits} + \text{Walks} * \Sigma\text{Bases}))/(\text{At-Bats} + \text{Walks})$$

The Runs Created formula reflected a team's ability to get on-base as a proportion of its opportunities through at-bats, hits and walks. James then evaluated historical baseball data using his model and found that Runs Created was a better model at predicting runs than other predictors (Lewis 2003). This formula better measured a player's offensive contribution than batting average, because wins are decided on the team with the highest number of runs, not the highest batting average. James further expanded upon and refined the RC formula to its current version.

$$\begin{aligned}RC = {} & ((H + BB + HBP - CS - GIDP) * (TB + 0.26(BB + HBP - IBB) \\ & + 0.52(SB + SH + SF)))/(AB + BB + HBP + SH + SF)\end{aligned}$$

In this formula, H is hits, BB is base-on-balls or walks, HBP is hit by pitch when a batter is struck by the ball during a pitch, CS is caught stealing or when the base runner is tagged out on base during a steal attempt, GIDP is grounded into double play or when the batter hits the ball on the ground and causes two outs, TB is total bases or number of bases reached safely, IBB is intentional base on balls or intentional walks when the pitcher purposely walks the batter, SB is stolen bases or the number of bases taken by the base runner between pitches, SH is sacrifice hit or when the batter hits the ball on the ground in such a way that an out will be created but a base runner will advance a base, SF or sacrifice fly is related to SH except the batted ball does not touch the ground.

Further instantiations of Runs Created led to Runs Created Above Average (RCAA) (Sinins 2007) which compares Runs Created to the league average (Woolner 2006) where positive values indicate a player performing above the league's average. Another derivative of RC is Runs Created per 27 Outs (RC/27) (James and Henzler 2002).

$$RC/27 = RC/(AB - H + CS + DP + SF + SH)$$

RC/27 is RC/(at-bats – hits + number of times caught stealing + number of times a player hits into a double play + sacrifice flies + sacrifice hits).

The RC/27 statistic models a player's complete offensive performance over the course of an entire game (27 outs). From further analysis, it was found that most bench players (i.e., players that do not start but come into the game later) will typically have 80% of the offensive capability of the starter. This 80% capability statistic varies for catchers at 85% and first basemen at 75% the starter's ability (Woolner 2006). It is believed that the performance differences between player positions ARE based on the physical demands of the position as well as how often the player is in the lineup. For instance, the catching position is more physically demanding than most other positions, including first base, which would necessitate the catcher to perform better than their teammates. Furthermore, certain pitchers will only work with a specific catcher, which allows multiple catchers to gain starter experience. The reduced capability of first baseman bench players is believed to be a byproduct of National League rules excluding the Designated Hitter (DH) position. This means that in the National League, pitchers are expected to also function as batters. Whereas in the American League, the DH position allows a substitute player to bat in the place of the pitcher, but not play defensively. These substitute players are oftentimes in the waning years of their careers and have lost some of their defensive ability due to age. Should these players be traded to the National League, they are still valued for their offensive capability, however, since they must play defensively under National League rules, these players are often put in the least physically demanding positions, typically first base.

3.3 Win Shares

In 2001s *Baseball Abstract*, Bill James introduced the concept of Win Shares, where players are assigned a portion of the win based upon their offensive and defensive input and further explained it in a follow-up book of the same name (James and Henzler 2002). Win Shares is a complicated formula that takes into account many constants and educated guesses, primarily because some of the measures were never captured in the historical data.

While still a matter of debate within the sabermetric community, Win Shares attempts to assign players credit for winning a game based upon their performance. Assuming a team has equal offensive and defensive capabilities, defense is credited with 52% of a win whereas offense is only 48%. This seemingly arbitrary division is justified as a way to even out the public's perception that offense is the more important component of a win. While the formula itself is still being refined within the sabermetric community, its results are difficult to argue with. Players with seasonal Win Shares of around 20 are typically all-stars, Win Shares of 30 indicates an MVP season and Win Shares of 40+ point to a historic season. For example, Barry Bonds had a Win Share of 54 in 2001 when he set the record of 73 homeruns in a single season.

3.4 Linear Weights and Total Player Rating

The Linear Weights formula calculates runs based upon the actions of the offensive player.

$$\begin{aligned}\text{Linear Weights} = {} & 0.47(1\text{B}) + 0.78(2\text{B}) + 1.09(3\text{B}) + 1.40(\text{HR}) + 0.33(\text{BB} \\ & + \text{HBP}) + 0.30(\text{SB}) - 0.60(\text{CS}) - 0.25(\text{AB} - \text{H}) \\ & - 0.5(\Sigma\text{Outs}_{\text{Base}})\end{aligned}$$

George Lindsey used this formula as an alternative to simple batting average (Albert 1997). Recognizing that there were three ways to get on base, hits resulting in a base (1B, 2B, 3B and HR), walks and being hit by a pitch (HBP), Lindsey further extended his model to reward those players that advanced through base stealing (SB), penalize players that were caught stealing (CS) and penalize those that were called out on the basepaths ($\Sigma\text{Outs}_{\text{Base}}$).

Pushing the idea of linear weights further, Total Player Rating (TPR) is a little more complicated and builds into itself comparisons for the position played and the ballpark (Schell 1999).

These comparisons allow statisticians to compare the performance of players to league averages based on the player's defensive position and the ballpark they are playing, because some ballparks may be more difficult for a position player than others. TPR has ratings of Batting Runs, Pitching Runs and Fielding Runs, which are (1) summed, (2) adjusted for player position and ballpark, and (3) divided by ten so that players can be compared against league averages. However, this statistic is also undergoing scrutiny where an average player is assumed to have a TPR of zero and Bill James claims the baseline should be measured elsewhere (James and Henzler 2002).

3.5 Pitching Measures

Pitching is another important staple of baseball and the performance of pitchers is closely watched by fans and sports organizations alike. Earned Run Average (ERA), which measures pitching performance over nine innings against the number of earned runs, runs that come from hits, is one of the most relied upon pitching statistics.

$$\text{ERA} = (\text{Earned Runs Allowed} * 9)/\text{IP}$$

Where IP is number of innings pitched. This statistic is usually coupled with a Win/Loss record and can be deceptive. Take for instance a poorly performing pitcher that plays on a team with a high powered offense. The pitcher will have a high ERA but also a deceptively high Win/Loss record. Similarly, an excellent pitcher playing on a team without run support will have a good ERA, but a poor Win/Loss record. In order to adjust to these situations, the Pitching Runs statistic was developed to more directly compare pitchers to league performance.

In Pitching Runs, the number of innings pitched is divided by nine innings then multiplied by the league's average ERA and then the earned runs allowed is subtracted out. The result of this formula gives the anticipated number of runs a pitcher would allow over the course of a complete game. Average pitchers would have a Pitching Runs score of zero while the Pitching Runs for better performing pitchers would be positive.

Another Bill James pitching measure is the Component ERA (ERC), which breaks out the different components of pitching outcomes and calculates them with the ERA.

$$\text{ERC} = (((\text{H} + \text{BB} + \text{HBP}) * 0.89(1.255(\text{H} - \text{HR}) + 4 * \text{HR}) + 0.56(\text{BB} + \text{HBP} - \text{IBB}))/(\text{BFP}*\text{IP}))*9 - 0.56$$

Where BFP is number of batters faced by the pitcher (Baseball Info Solutions 2003). However, the ERC goes into more complicated formulation under certain conditions and other organizations offer differing models.

4 Basketball Research

APBRmetrics aims to capture basketball statistics in terms of team rather than individual performance.

Basketball experienced its own sabermetric revolution shortly after baseball's (Pelton 2005). With their own wealth and depth of statistics, several pioneers of basketball statistics set about to better quantify and assign credit through the creation of APBRmetrics, named for the Association of Professional Basketball Researchers (ABPR). APBRmetrics is fundamentally different from sabermetrics, because APBRmetrics attempts to view statistics in terms of team rather than individual performance. One such example of this is team possession and how effective the team is at scoring points. The thinking is that since teams must function as cohesive units, they should be analyzed as units rather than attempting to quantify team chemistry, or how well players perform with one another, at the individual level.

To back up this point, players can have either a positive or negative impact on team performance. As an example, during the 2004–2005 season, Stephon Marbury had a −0.4 point negative impact on the team while he was on the court. At the outset, this statistic would indicate that Marbury was performing at or slightly below average. However, when Marbury was off the court and not helping his team, the team had a −12.0 point deficit. This 11.6 point differential when Marbury was on the court versus off the court, illustrates that Marbury can best improve his team's performance when he is on the court.

4.1 Shot Zones

A basketball court can be divided up into 16 areas where a player on offense might be inclined to shoot a basket. By analyzing the percentage of player success from each of these zones, defensive adjustments can be made to limit scoring while offensively, coaches may try to maximize these types of shots (Beech 2008b). Figure 4.1 illustrates the different shot zone locations.

From a Shot Zone analysis on the 2004–2005 season, 82games.com found that for three point shots from the corner (Zones 1 and 5), Golden State's Michael Dunleavy had the highest accuracy of 0.571 from the left corner and Sacramento's Cuttino Mobley had 0.600 accuracy from the right corner. Likewise, shot zones can portray player tendencies. For example, under the basket (Zones 13 and 14), Miami's Shaquille O'Neal made the most attempts in the league but was not very

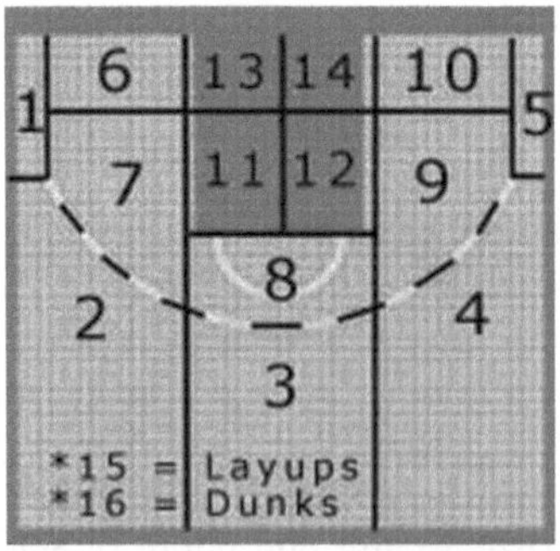

Fig. 4.1 Shot zone layout (82games.com)

successful, 0.416 from the left and 0.424 from the right. Knowing where players are successful and comfortable can lead to better offensive or defensive strategies.

4.2 *Player Efficiency Rating*

Player Efficiency Rating (PER) is a complex per-minute rating of player effectiveness that rewards positive contribution and penalizes negative ones (Hollinger 2002). This formula takes on many variables including assists, blocked shots, fouls, free throws, made shots, missed shots, rebounds, steals and turnovers to quantify player performance in regards to their pace throughout the game and the average performance level of the league.

PER is still a matter of debate and Hollinger admits that it does not take into account all of performance related criteria, such as hustle and desire (Hollinger 2002).

4.3 *Plus/Minus Rating*

Another method of calculating performance is through the Plus/Minus Rating system where each player is evaluated by calculating the number of points the team makes with that player on the field minus the number of points the opposing team receives. This calculation is done for each team player while they are on court and while they are on the bench. Player contribution can then be measured as the differential between their on and off court presence (Rosenbaum 2004).

For example, if Lebron James starts a game, (e.g., the score is 0–0) and then exits the game at 102 to 80, this +22 point difference would be Lebron's plus/minus rating for the game. Likewise, if John Stockton entered a game at 80 to 78, his team up by 2, and then leaves the game at 80 to 94, his team down by 14, Stockton would receive a −16 plus/minus rating.

Positive values indicate the player is making a positive point-based contribution to the team whereas negative values would point towards detrimental activity. Take for example Dwight Howard during the 2004–2005 season. Howard had a plus/minus rating of -2 when he was on the court versus an even rating when he was not (Rosenbaum 2005). This would seem to indicate that the team was better off without Howard's presence. However, the plus/minus rating system is not without its own share of quirks. Critics of the system point to its over-valuing of players that take a high number of shots and commit a large number of turnovers, which is not a beneficial team activity.

4.4 Measuring Player Contribution to Winning

A further metric to evaluate player contribution versus a substitute player is to adjust the plus/minus rating system to account for the talent level of teammates (Rosenbaum 2004). The reasoning is that player performance does not occur within a vacuum, but rather is a function of the overall team effort. The adjusted plus/minus is a regression estimate where the constant is the home court advantage against all teams, the kth order constants are the plus/minus differences between player K and the players of interest, holding all others constant. The x values, x1 through x14 refer to game level statistics per 40 min of play: points, field goal attempts at home, field goal attempts on the road, three point attempts, free throw attempts, assists, offensive rebounds, defensive rebounds, turnovers, steals, blocks, personal fouls, $(\text{points} * \text{assists} * \text{rebounds})^{1/3}$ and minutes per game. Regressing these 14 values together nets the adjusted plus/minus rating.

4.5 Rating Clutch Performances

The reason that 40 min is typically studied rather than the standard game time of 48 min, is the belief that the final minutes of the game are completely different from the rest of the game (Ilardi 2007). In instances where one team is ahead by several points, the lagging team may participate in fouling in order to retrieve possession of the ball. This behavior tends to skew these statistics from normal game behavior. The final 8 min of game time and any overtime, if necessary, provided that the scores of the two teams are within five points of one another, is referred to as the clutch. Some players tend to excel during this period, leading APBRmetricans to study the clutch performances of players. Some would argue that player contribution during the clutch is more important than during the rest of regulation play, because the prospect of winning or losing hangs in the balance. Others point towards legends of the game that have defined themselves with clutch performances (e.g., Bill Walton and Michael Jordan). Using PER during this period can identify new insights into offensive capability. To test defensive match-ups in man to man coverage, its assumed that one player's PER (i.e., looking at PER's of opposing positions), will be superior to that of their counterpart if a clutch performance is occurring.

While PER is limited to man to man coverage and cannot be used in other types of defenses such as zone (where the defensive player is confined to a specific area of space), this method can still provide valuable insight into player execution.

Another strategy for measuring clutch performance is to evaluate the performance of the team as a whole by using the plus/minus rating system and aggregating clutch points between players on the court during the last 8 min of the game. This provides additional insight into a player's clutch abilities by showing both on-court and off-court results in terms of point contribution.

5 Football Research

Advances in the statistical techniques of American football have not reached the levels of those collected in both baseball and basketball. For this reason, there is a lack of certain statistical data on individual players. While some basic statistics are collected such as number of touchdowns, receptions and interceptions, these aggregate counts pale in comparison to their sabermetric counterparts. Another reason football lacks data is because of the number of games played. The NFL plays 16 regular-season games compared to baseball's 162 and basketball's 82 games. Despite the lack of data and depth, there are several metrics meant to bridge these deficiencies.

5.1 Defense-Adjusted Value Over Average

The Defense-Adjusted Value Over Average (DVOA) is a comparative measure of success for a particular play (Schatz 2006). This statistic treats each play as a new event and measures the potential of success versus the average success of the league. Certain variables are taken into account such as time remaining, the down, distance to the next down, field position, score and quality of opponent. These variables carry different rewards if met and can be used to measure a particular player's contribution or aggregated to highlight team-based performance. A DVOA of 0% indicates that the defense is performing on par with the league average. Whereas positive and negative DVOA values indicate that the defense is performing above or below league averages respectively.

In football, possession is broken into four downs (or plays) with a sub-goal of exceeding a set number of yards before the expiration of downs. DVOA considers that in order to meet this sub-goal, 45% of the required yards should be gained on the first down, 60% on the second down and 100% by the third or fourth down (Carroll et al. 1998). If the play is deemed to be successful DVOA assigns it one point. If the play is successful early (e.g., attaining the sub-goal in the early downs), more points are awarded.

5.2 *Defense-Adjusted Points Above Replacement*

Defense-Adjusted Points Above Replacement (DPAR) is a player-based statistic that is compiled over the course of a season (Schatz 2006). DPAR is used to determine the point-based contribution of a player as compared to the performance of a replacement player. If a player is said to have a +2.7 DPAR, it means that the team should score 2.7 points because of the player's presence in the lineup, whereas that 2.7 points would be lost if the player was substituted by an average replacement.

5.3 *Adjusted Line Yards*

Adjusted Line Yards (ALY) is a statistic to assign credit to an offensive line based on how far the ball is carried (Schatz 2006). This statistic attempts to separate the running back from the contribution of the offensive line and is measured against league averages. If a running back is brought down behind the line of scrimmage (i.e., takes a loss of yards), the offensive line will be penalized heavily for the failure. If the same running back manages to make a long gain (i.e., picked up many more yards than usual), the offensive line is given minimal credit, because the offensive line can only make so much of a contribution and much of it is up to the running back. The ALY is further adjusted to league averages.

6 Emerging Research in Other Sports

> Aside from professional baseball, football and basketball, many other sports such as: NCAA football and basketball, soccer, cricket, and Olympic curling, are experiencing their own statistical awakening.

Aside from professional baseball, football and basketball, many other sports are experiencing their own statistical awakening. In soccer there is pioneering work in predicting the likelihood of injury based on biomedical monitoring as well as isolating the features that lead to tournament wins. NCAA College Basketball researchers can predict tournament matchups and victories with impressive accuracy. Two other sports, Olympic Curling and Cricket, are similarly gathering data on their opponents and analyzing the factors that contribute to winning. It is not a far stretch to adapt any these techniques to other sports.

6.1 NCAA Bowl Championship Series

NCAA Football uses data mining and knowledge management techniques to rank collegiate teams. Because NCAA football does not enter into a tournament style of play like basketball does, disputes routinely break out regarding which two teams should compete for the National Championship. The Bowl Championship Series or BCS, was created to address these problems, however, it became a part of the controversy in 2004 when the University of Southern California (USC) was rated number one by the Associated Press poll and number three by the BCS. As the post-season ended, both LSU and USC were crowned co-champions. Following 2004, the BCS algorithm was rewritten. However, this rewritten algorithm led to further disputes in 2005 when Auburn was ranked number three in the BCS and was the only undefeated team that year. Even further problems emerged in 2007 when Boise State went undefeated yet was ranked #5 and in 2008 when Utah went undefeated yet was not allowed to play for the national championship.

The BCS is a composite type algorithm, in which many various polls are taken into account and weighted accordingly. In particular, the BCS uses the Harris Interactive College Football poll, the Coaches poll (what rankings fellow football coaches believe is fair) and computer polls including Jeff Sagarin's NCAA football poll at USA Today and the Seattle Times. Each team is then assigned points based upon their poll ranking in all of the component polls. Teams are then rank ordered based on their score.

6.2 NCAA Men's Basketball Tournament

NCAA basketball has its own share of research. One notable figure is Jeff Sagarin, who publishes his basketball rating system based on a team's win/loss record and the strength of their schedule (USA Today 2008). However, more research exists concerning the NCAA Men's Basketball Tournament. Every March, college basketball enters into March Madness – a tournament where 64 Division I teams will compete for the title of National Champion. While the exact selection process for the 64 teams is not made public, a Selection Committee makes the determinations and the 64 teams are selected on a "Dance Card." Two researchers interested in this process, developed a method of predicting the at-large bids with a 93.0% success rate over the past 16 years (Coleman and Lynch 2009). This would seem to indicate that the Selection Committee uses similar selection techniques every year, even though the membership of the committee changes from year to year (SAS 2005). The technique weights 42 pieces of information on each team, including their RPI ranking (or relative strength against other teams), win/loss record, conference win/loss record, etc. and forms a rank order score called the "Dance card score" (Coleman and Lynch 2001).

Once teams have been selected for the playoffs, the same team of researchers has devised a second algorithm, "Score Card," to predict the winners (Coleman and Lynch 2008). Using data from the 2007 tournament, their system was able to correctly predict the winners for 51 of the 64 games, an accuracy of 79.7%. The Score Card algorithm is remarkably simpler than its counterpart Dance Card, because only four variables are necessary; the team's RPI value, RPI value of the team against non-conference opponents, whether the team won the conference title and the number of wins in their previous ten games.

6.3 Soccer

Soccer arguably garners the most passionate fans worldwide. With such devotion to the sport, it is understandable that many researchers and fans alike have an interest in predicting prestigious tournament outcomes. While one such study found that time of possession is an important factor in game outcomes (Papahristoulou 2006), other studies have noted that country of origin and home field advantage were sizable factors in predicting team success (Barros and Leach 2006). From this later study of the teams comprising the UEFA Tournament, researchers used a myriad of factors including league win/loss records, tournament win/loss, shots, team record at home and on the road and past tournament performance to predict not only who the strongest teams will be, but also to forecast which team should win the tournament.

Another important contribution is the ability to forecast when a player may be experiencing the onset of an athletic impairment through injury prediction. Oftentimes a player, regardless of their sport, will try to play despite their injury or performance degradation. AC Milan has been piloting predictive software that monitors the workouts of their players (Flinders 2002). This software compares an athlete's workout performance against that of a baseline, and any drops in performance may indicate that the player may be injured. Other biomedical methods employ a series of weighted variables including injury rate, odds of injury and history of injury to compile a risk likelihood measure (Hopkins et al. 2007). Another method looks at 17 various risk factors, such as previous injuries, playing characteristics, endurance and game-time preparation among others, and it was found that inadequate warm-ups were the usual factor in injury-related events (Dvorak et al. 2000).

6.4 Cricket

Similar to the wealth of statistics kept in baseball, the game of Cricket also holds an extensive store of data within the Wisden Almanack, going back to 1864 (CricInfo

2008). This data has also been recently explored using data mining and knowledge management tools to some success. In a study of One Day Test Cricket matches, it was found that a mix of left/right batsmen and a high runs to overs ratio were both highly correlated to winning (Allsopp and Clarke 2004). The usage of alternating left and right-handed batsmen is believed to keep the opposing team's bowler out of their typical rhythm and thus be less effective (Allsopp and Clarke 2004). The high number of runs to overs ratio, (e.g., amount of runs scored as a proportion to the number of offensive periods) indicted that a quicker paced game (i.e., more runs) was also a factor in determining a winning team. These factors can further be used to determine team effectiveness and also tournament play.

6.5 Olympic Curling

The Curling event in the 1998 Winter Olympics would appear to be an atypical place to find data mining and knowledge management tools at work. During the 8 days of Curling competition at the Nagano Olympics, IBM collected a wealth of data on players, strategy, the precise paths the stones took as well as event outcomes (Taggart 1998). While this data collection was not extensively used at the time, the potential still exists to isolate a Curling player's tendencies and weaknesses (Cox and Stasko 2002).

7 Conclusions

With many sports using some form of statistics to measure player performance, team cohesiveness, injury probability and their opposition's trends and tendencies; statistics have become a valuable tool in sports. Using these tools and extrapolating trends from them is the next step in knowledge creation.

8 Questions for Discussion

1. What statistics are considered more important than others in a particular sport? What are some of their strengths and weaknesses?
2. What sports could benefit from a closer examination of performance measurement and why?
3. Is there anything you would improve in existing statistics for different sports?

Chapter 5
Tools and Systems for Sports Data Analysis

Chapter Overview

This chapter investigates some of the data mining and scouting tools available for sports analysis. In particular, we analyze the role of these tools and how they can help an organization. Tools such as Advanced Scout, which maintains play-by-play data in an easy to query environment and Inside Edge, which provides pictorial descriptions of player tendencies, will be investigated. Sports fraud detection is another interesting area where sport-related data can be analyzed against historical patterns to identify potential instances of sports fraud from players, corrupt officials or even suspicious bettors.

1 Introduction

Systematic data mining and knowledge management tools have been mainly constrained to in-house analyses by sports organizations. However, simpler tools using the theories of Bill James and his contemporaries made their debuts with fantasy team managers and rotisserie leagues before Lewis' Moneyball revolution. These early adopters found success by using data mining and knowledge management tools which further led to the development of advanced measurement techniques and more knowledge-based systems. There is a growing market for third-party vendors that harness data mining and knowledge management tools to sell niche services to individuals and sporting organizations. These niche services can serve a variety of areas including isolating player tendencies, providing more in-depth scouting reports and uncovering instances of fraudulent activity within the sports environment.

R.P. Schumaker et al., *Sports Data Mining*, Integrated Series in Information Systems 26, DOI 10.1007/978-1-4419-6730-5_5, © Springer Science+Business Media, LLC 2010

2 Sports Data Mining Tools

Sports data mining and scouting tools such as: Advanced Scout, Digital Scout, Inside Edge, and Sports Data Hub, are gaining popularity for sports analysis.

One area of third-party development has been the design of tools that do not fit the traditional notion of data mining. One such instance is to incorporate elements of game video footage that can be drilled down into its component pieces and queried. Virtual Gold is an example of such a company that provides this type of unique service. Other distinctive methods include simple graphical analysis of existing statistics, allowing domain experts to more readily identify patterns within the data. Information visualization has long been recognized as an effective tool for knowledge management (Zhu and Chen 2005).

2.1 Advanced Scout

Advanced Scout was developed by IBM during the mid 1990s as a data mining and knowledge management computer program. Its purpose is to glean hidden patterns within NBA game data and provide additional insights to coaches and other organization officials. Advanced Scout not only collects structured game-based statistics during play, but also unstructured multimedia footage. With the entire NBA league having access to Advanced Scout, coaches and players can use this tool to prepare for upcoming opponents and study them using historical footage (Shulman 1996).

The multimedia aspect of Advanced Scout collects raw game-time footage, error-checks the content and finally segments it into a series of time-stamped events such as shots, rebounds, steals, etc. (Bhandari et al. 1997). The processing and error-checking stage is a rule-based series of procedures to verify the consistency and accuracy of the data. The error-checking process includes removing impossible events (events tagged incorrectly), looking for missing events and attributing plays to particular players. In cases where the rule-based strategy is unable to identify key elements, a domain-expert can manually label events in the game footage.

Advanced Scout also possesses a knowledge management component called Attribute Focusing, where a particular attribute can be evaluated over the entire distribution of data and display the results in both textual and graphical descriptions of the anomalous subsets. Those subsets with a distinctly different statistical distribution are then set aside for further analysis by players or coaches (Bhandari 1995). For example, consider the following textual description from Advanced Scout:

> When Price was Point-Guard, J. Williams missed 0% (0) of his jump field-goal-attempts and made 100% (4) of his jump field-goal-attempts. The total number of such field-goal

attempts was 4. This is a different pattern than the norm which shows that: Cavaliers players missed 50.70% of their total field-goal-attempts. Cavaliers players scored 49.30% of their total field-goal-attempts (Bhandari et al. 1997).

This description illustrates an easy to read analysis of the anomalous behavior of Williams when Mark Price was the Cavaliers point guard. Once a coach or player receives this information, it is up to them to determine why this is the case. For the above example, it was determined that when Price was double-teamed, he would pass the ball to Williams for wide-open jump-shots.

Aside from the anomaly detection facet of Attribute Focusing, Advanced Scout can be queried to find a relevant game-time event such as particular shots, rebounds, etc. Players and coaches alike can use this information to hone skills and better understand player dynamics.

2.2 *Synergy Online*

Similar to Advanced Scout is Synergy Sports Technology's Synergy Online. This product is dedicated to basketball-based multimedia and contains an index of live video broadcasts as searchable media. With this system, coaches, players and fans can query plays in real-time and receive constantly updating player statistics. Synergy Online also allows for live broadcasts to stream to desktops or mobile devices. This same company is also the driving force behind gaming console products such as NBA Live 09 and NBA Live 10, by providing real game data to gamers. Their signature product, Digital DNA, models nearly 1,000 characteristics and tendencies of a player (Colston 2009) in order to better model their expected performance and bring a realism to the gaming experience in real-time (Arnovitz 2009).

2.3 *SportsVis*

Another way of finding interesting data patterns is to identify them graphically. SportsVis is such a tool that allows users to view a plethora of data over a selected period of time (Cox and Stasko 2002). This data could include team runs over an entire season or player-specific criteria such as the runs scored off professional baseball pitcher Curt Schilling over a 32 game period, as shown in Fig. 5.1.

This graphical description may indicate trends or uncover potential problems such as injuries. Other interesting visualization techniques can be found in Baseball Hacks, where author Joseph Adler walks users through the process of using Excel and Access databases to view various baseball statistics (Adler 2006). These techniques include batter spray diagrams where a hitter may favor hitting the ball to certain portions of the field under certain situations, and frequency distributions using many of the sabermetric statistics.

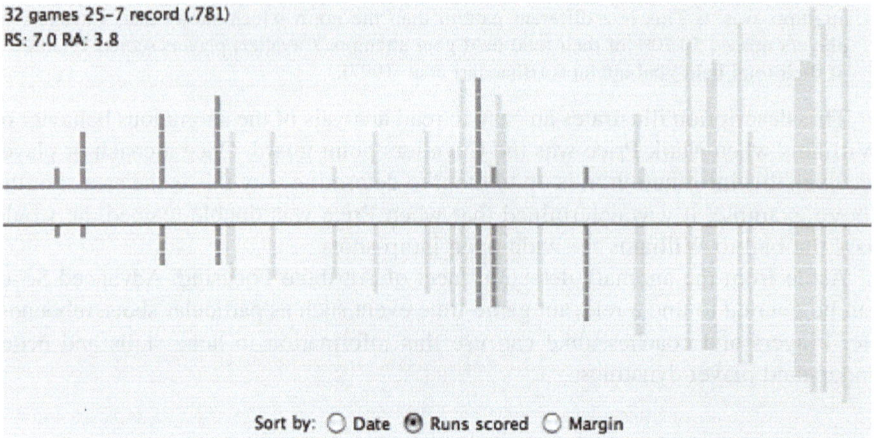

Fig. 5.1 Curt Schilling runs scored over 32 games (Cox and Stasko 2002)

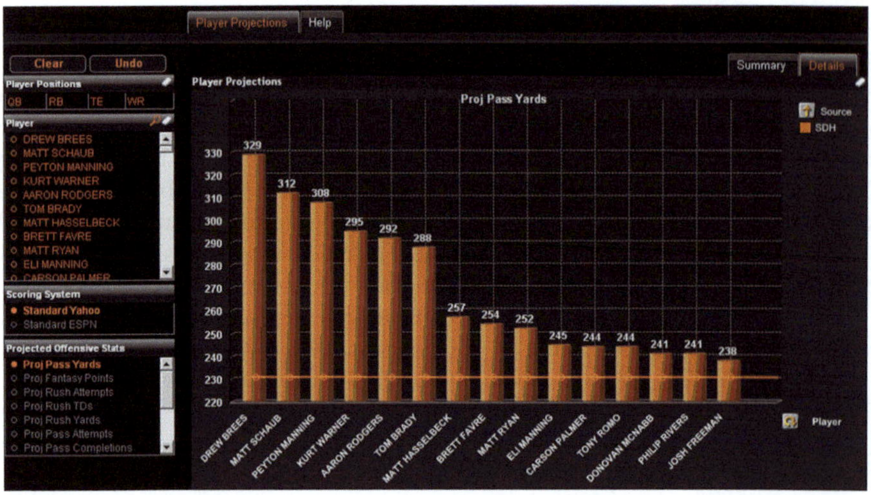

Fig. 5.2 Projected passing yards, courtesy of http://analyze.sportsdatahub.com/projections/basic/PlayerProjections.aspx

2.4 Sports Data Hub

Sports Data Hub (http://ww2.sportsdatahub.com) is an interactive data source for fantasy team owners. With its fascinating and seemingly endless array of graphical data, fantasy team owners can either choose to immerse themselves in reams of data, or use a shortcut tool that can project which players will score the most fantasy points using either Yahoo's or ESPN's fantasy point ranking system. Figure 5.2 shows one such graphic of projected passing yards for NFL quarterbacks.

3 Scouting Tools

Scouts used to rely on manual methods to keep track of player performance. Today that power is placed in the hands of fans and the next generation of scouts. Game statistics can be input on the fly during the game and complete reports of overall play as well as individual attributes can be disseminated for player improvement.

3.1 Digital Scout

Digital Scout is the computerized answer to inputting statistics and creating score cards. Fans and sports organizations alike can use this software on a palmtop, laptop or desktop machine to collect and analyze game-based statistics. This software can be adapted for all major sports that utilize some form of statistical record-keeping. Digital Scout can also allow users to print score results or create custom reports on particular attributes, such as baseball hit charts, basketball shots and football formation strengths (Digital Scout 2008).

This software has been found to be very useful and has been adopted by Baseball's Team USA (Petro 2001), Little League Baseball (Petro 2003), and certain basketball tournaments (Weeks 2006).

3.2 Inside Edge

Another scouting tool is Inside Edge which was created by Randy Istre and Jay Donchetz in 1984. Inside Edge provides pitch charting and hitting zone statistics for college and professional baseball teams (Inside Edge 2008a). Coupled with a professional scouting department, Inside Edge has been used by many MLB ballclubs including all of the World Series champions between 1996 and 2001. The strength of Inside Edge is in easy to read visual scouting reports that employ a host of textual and graphical elements as well as expected opponent strengths, weaknesses and tendencies.

Reports on strengths, weaknesses and tendencies are all backed by statistical data and users can examine this data in detail. An example spray chart of Rafael Furcal of the Atlanta Braves is shown in Fig. 5.3. Note the density of infield hits shown at second base. As an opposing team, the second baseman should expect a higher proportion of Furcal's hits to be grounded towards him.

Another more complete report is the Pitcher Postgame as shown in Fig. 5.4. You can easily see the increases in the velocity of pitches as the game progresses (e.g., from 92 to 95 mph for fastballs) as well as pitch effectiveness (e.g., opposing left-handed batters, LHBs, perform poorly against Bartolo Colon's fastball pitch with a 0.167 batting average).

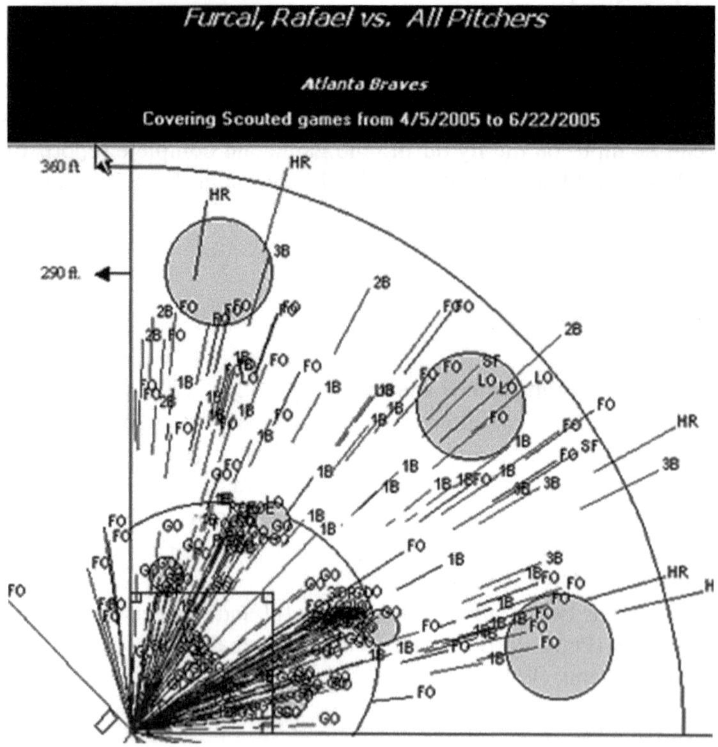

Fig. 5.3 Spray report of Rafael Furcal (Inside Edge 2008b)

The graphical representation of pitcher performance in the strike zone, based on individual statistical performance, can allow pitchers to visually understand the areas of the strike zone that they are most effective.

4 Sports Fraud Detection

Sports fraud detection using data mining techniques can help keep games in check. Organizations such as Las Vegas Sports Consultants Inc (LVSC) regularly analyze both betting lines and player performance, looking for any unusual activity.

Fraud in sports has always been a problem. Some scandals have led to historical precedence, e.g., the eight players from the 1919 Chicago White Sox that threw the World Series and resulted in those players receiving bans from organized baseball for

Pitcher Postgame Report
Colon, Bartolo - Angels

Inside Edge SCOUTING SERVICES

Game Information / Totals

Date: 6/15/05 **Opponent:** Nationals **Location:** Anaheim

IP	TPs	PA	AB	H	2B	3B	HR	KS	KC	BB	IBB	HBP	Pit's/PA
9.0	92	34	33	7	0	0	1	1	1	0	0	0	2.7

Average Velocity by Pitch Count

	Avg.	Low	High	Pit's 1-15	Pit's 16-30	Pit's 31-45	Pit's 46-60	Pit's 61-75	Pit's 76-90	Pit's 91-105	Pit's 106-120	Pit's 121+
Fastball	93	89	96	92	93	94	92	95	92	95	0	0
Curve Ball	0	0	0	0	0	0	0	0	0	0	0	0
Slider	85	82	89	84	84	86	84	88	0	0	0	0
Changeup	84	83	86	0	83	84	84	84	86	0	0	0
Other	0	0	0	0	0	0	0	0	0	0	0	0

Pitch Breakdown vs. RHBs

				1st Pitch		
PA : 12	% (#) Pit's	Str. %	Opp. BA	% (#) Pit's	Str. %	Opp. BA
Fastball	83% (24)	79%	.200 (2/10)	83% (10)	90%	.000 (0/3)
Curve Ball	0% (0)		(0/0)	0% (0)		.000 (0/0)
Slider	17% (5)	60%	.500 (1/2)	17% (2)	100%	.500 (1/2)
Changeup	0% (0)		(0/0)	0% (0)		.000 (0/0)
Other	0% (0)		(0/0)	0% (0)		.000 (0/0)
Total	100% (29)	76%	.250 (3/12)	100% (12)	92%	.200 (1/5)

Pitch Breakdown vs. LHBs

				1st Pitch		
PA: 22	% (#) Pit's	Str. %	Opp. BA	% (#) Pit's	Str. %	Opp. BA
Fastball	87% (55)	76%	.167 (3/18)	91% (20)	60%	.250 (1/4)
Curve Ball	0% (0)		(0/0)	0% (0)		.000 (0/0)
Slider	5% (3)	100%	.000 (0/1)	0% (0)		.000 (0/0)
Changeup	8% (5)	80%	.500 (1/2)	9% (2)	50%	.000 (0/1)
Other	0% (0)		(0/0)	0% (0)		.000 (0/0)
Total	100% (63)	78%	.190 (4/21)	100% (22)	59%	.200 (1/5)

NOTE: Total pitches include unspecified pitch types.

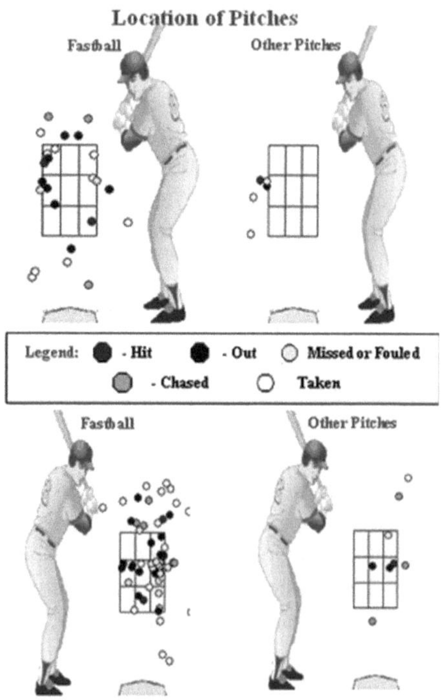

Fig. 5.4 Pitcher postgame report for Bartolo Colon (Inside Edge 2008b)

life. Some scandals led to controversy, e.g., baseball's Pete Rose's alleged betting on the Cincinnati Reds, the team he was managing at the time, which led him being placed on baseball's ineligible list. Other scandals are more widespread such as the use of performance-enhancing drugs in sports. When fraudulent activity in sports occurs, it generally falls into one of three categories: poor player performance, a pattern of unusual calls from the referee, and lopsided wagering (Audi and Thompson 2007).

Poor player performance, or point shaving, is one way in which game integrity can be compromised. This involves a player or group of players that purposefully underperform in order to affect the game's betting line. Before two teams physically meet for a match, sportsbooks set a betting line which will draw an equal dollar amount of wagers for either team, that way the losing side of the wager pays the winning side minus the sportsbook's commission. Should the line become unbalanced, the sportsbooks would be responsible for the difference and would either cause them to either lose money, lose business, or both. If one team is heavily favored, then the line will be more distinct with the heavily favored team having to achieve a larger victory in order to win the wager. Point shaving is simply a player attempting to manipulate the outcome of the game by not meeting the betting line. A recent study into NCAA basketball, found that 1% of games involve some form of point shaving (Wolfers 2006). Being able to discover instances of point shaving is incredibly difficult (Dobra et al. 1990), especially when there is typically no serial correlation in betting markets from game to game (Oorlog 1995).

A pattern of unusual calls from the referee can also influence game outcome. Similar to point shaving, compromised referees can also manipulate the betting line. Referees have it in their power to make the game easier or harder for a team, and thus influence the betting line (Igloo Dreams 2007). A recent example of this was in the summer of 2007 when NBA referee Tim Donaghy was investigated and convicted for compromising basketball games to pay off gambling debts.

Both point shaving and questionable referee calls have the same outcome in mind, making money. Thus lopsided wagering can be used as an indicator of a compromised game. This type of wagering could involve betting in excess of what is normally expected or betting heavily against the favorite. In one particular example, a gambler from Detroit made repeated bets against the University of Toledo versus Temple in college football (Audi and Thompson 2007). One of the wagers, $20,000, was four times larger than what was considered to be a typical large wager for that conference. This gambler correctly picked that Toledo would be unable to achieve a required number of points and suspicions were raised from Sportsbook operators. In the following game, more atypical wagers began coming in against Toledo forcing one of the Sportsbooks to cancel Toledo events from their boards. Sportsbooks make their money through evenly positioning the wagers, one side hands their money over to the other, minus a commission. When games are compromised and the wagers uneven, the Sportsbook will lose money on the event. So it happens to be in the Sportsbooks best interest to keep integrity within sports and to set unbiased betting lines (Paul and Weinbach 2005).

4.1 Las Vegas Sports Consultants

One of the organizations that actively looks for fraudulent sports activity is Las Vegas Sports Consultants Inc (LVSC). This group sets betting lines for 90% of Las

Vegas casinos. The LVSC statistically analyzes both betting lines and player performance, looking for any unusual activity. Player performance is judged on a letter-grade scale (i.e., A–F) and takes into account variables such as weather, luck and player health (Audi and Thompson 2007). Taken together, games are rated on a 15 point scale of seriousness. A game rated at 4 or 5 points may undergo an in-house review, 8–9 point games will involve contact with the responsible league. Leagues are similarly eager to use the services of LVSC to maintain game honesty. The LVSC counts several NCAA conferences, the NBA, NFL, and NHL, as some of its clients.

4.2 *Offshore Gaming*

Las Vegas Sportbooks are not the only gambling institutions with an interest in honest and fair sports events, offshore betting operations arestarting to fill this role as well. One popular offshore gambling site, Betfair.com, has signed an agreement with the Union of European Football Associations (UEFA) to help monitor games for match-fixing, unusual results, or suspicious activity (Jimmy 2007). Betfair is an internet betting exchange, a gambler's eBay, which pairs up via the web those who want to have a bet with those who want to take one, cutting out the traditional middleman (bookmaker) (Cameron 2008). The company has become one of the world's ten biggest companies operating on the web and has contributed to exponential growth in internet betting. While difficult to detect all instances of gambler-tainted matches, these beginning steps can assure fans and bettors alike that vigilance is taking place and that player and referee activity is being scrutinized.

5 Conclusions

Data mining tools for players, scouts and interested third-parties have become more commonplace and almost a requirement in certain circles. These tools all vary in their size, scope and level of detail. From simple textual outputs of anomalous game behavior to elaborate visualizations that allow individuals quick and easy uptake of complex data, these tools have advanced new ways and ideas to think about sports. Aside from their ability to provide a competitive advantage, data mining tools can help keep games in check through constantly monitoring game data for suspicious patterns.

6 Questions for Discussion

1. What other sport-related data mining tools do you have experience using?
2. Data mining tools are mostly sport independent. Do you agree or disagree with this assessment and why.
3. What other areas of sport-related fraud could be analyzed using data mining?

Chapter 6
Predictive Modeling for Sports and Gaming

Chapter Overview

Predictive modeling has long been the goal of many individuals and organizations. This science has many techniques, with simulation and machine learning at its heart. Simulations such as basketball's BBall can model an entire season and can deduce optimal substitution patterns and scoring potential of players. Should unforeseen events occur such as an unexpected trade or long-term injury, additional simulations can be performed to assess new forms of action. Aside from the potential of simulations, machine learning techniques can uncover hidden data trends. Greyhound racing is one such area that has been explored with many different machine learners. While the choice of algorithms used in each study may differ, they all had one common similarity, they beat the choices human track experts made and were able to use the data to create arbitrage opportunities.

1 Introduction

It has long been the goal of many researchers and gamblers to correctly predict outcomes based upon historical evidence. The motivation for doing so has led to many sport-specific developments such as statistical simulations and sport-independent techniques in machine learning. Using these tools, trends in sport data can be identified and manipulated for personal, competitive, or economic advantages.

Within this predictive modeling domain, one area has received considerable attention, not only from an academic statistical perspective, but also from anecdotal evidence from the sports environment which oftentimes lends credence where none is needed. The concept of streaky player performance has been well studied and is also known as the "hot-hand" effect, where performance is elevated above average for an extended period of time. In research on the hot-hand effect in basketball, researchers argued that if the hot-hand effect existed, making a shot would increase the chance of

R.P. Schumaker et al., *Sports Data Mining*, Integrated Series in Information Systems 26, DOI 10.1007/978-1-4419-6730-5_6, © Springer Science+Business Media, LLC 2010

making another shot (Tversky and Gilovich 2004). However, from empirical study, the success of a shot was found to be independent of previous shot outcomes.

The same streaky behavior concept applies itself to flipping a coin or landing on a specific color on a roulette wheel. In flipping a fair coin, a balanced coin that displays either heads or tails with equal proportion, the coin can be expected to exhibit an equal number of heads and tails. Further, each flip can be considered independent of every other flip. The coin does not maintain a memory of its prior flips to influence the current one. So a coin that has a history of one heads flip will have the same chance of landing on tails as one with six consecutive heads flips. While these events may be independent of one another, oftentimes people will try to ascribe a pattern to this type of random behavior. The same behavior can be observed in a fair game of roulette, where the roulette ball will land on either the red or black square with equal frequency and independent of prior activity. However, humans will often believe in a streak where none exists.

However, the players in a sporting event can hardly be considered balanced, fair, and independent of their prior actions. This is where psychology begins to take over, where streaks and streaking behavior are possible, but not well understood.

Baseball still has its adherents to streaky behavior. In an interesting piece of research that sought to model streaky player performance, it was found that there are certain players that exhibit significant streakiness, more than what probability can account for (Albert 2008). This is where simulation and machine learning begin to become important tools to isolate and tease out instances of these behaviors, so that predictions can be made.

2 Statistical Simulations

> Statistical simulations involve the imitation of new game data using historical data as a reference. Simulations can be performed in a wide variety of sports domains including baseball, basketball, football, and hockey.

Statistical simulations involve the imitation of new game data using historical data as a reference. Once this imitation data has been constructed, it can be compared against actual game play to test the accuracy of its predictive power. Simulations can be performed in a wide variety of sports domains including baseball, basketball, football, and hockey.

2.1 Baseball

Baseball has long been a hotbed of simulation, with fantasy and rotisserie leagues to name two. Simulations can be made on finding the optimal pinch hitters using

Markov chains, where matrices of players, inning states (top or bottom of the inning), number of outs and the on-base possibilities are all taken into account and multiplied by substitution matrices using pinch hitters (Hirotsu and Wright 2003). This method can then be used to find the optimal pattern of player substitutions based upon the given situation.

A player-focused simulation method, developed at Loyola Marymount, uses historical player data to predict future homerun totals by analyzing the frequency distributions of homeruns, where top performances (i.e., record-breaking seasons) are considered "large" events and then relating those large event frequencies to the frequencies of smaller events (i.e., individual homeruns) (Kelley et al. 2006). This research is based on the similar approach used to model earthquake frequency and intensity distributions. Applied to baseball, a solitary hit can be described as a small tremor, whereas a cluster of hits would increase the model's intensity. Further taking this model and extrapolating it for a specific player, predictions of their on-going batting tendencies can be made. To put it loosely, if the ball is flying out of the park more than usual during a season, the potential exists for someone to have a terrific year, leading to the observation that such historical performances are often linked.

Another study that investigated the process of predicting baseball's division winners, those that finish first within their respective division, used a two-stage Bayesian model based on a team's relative strength, measured by winning percentage, batting averages, the ERA of the starting pitcher and home field advantage, where it is suspected that teams playing at home possess an advantage (Yang and Swartz 2004). This study simulated MLB baseball's entire 2001 season and their method was found to be surprisingly accurate in predicting five of the six division winners by July 30th. Other Bayesian models such as predicting Cy Young winners (best pitcher in the league that season) have also netted similar accuracy results (Smith et al. 2007).

2.2 *Basketball's BBall*

In basketball, one popular simulation tool is BBall. It was developed by basketball researcher Bob Chaikin, a consultant of the Miami Heat (Solieman 2006). This software uses historical data and APBRmetrics to simulate anywhere from one game to an entire season. Developed for NBA coaches, scouts and general managers, BBall can determine a team's optimum substitution pattern over the course of a season (e.g., the pattern that produces the most simulated wins), the effect a player trade may have on the team's performance, the effect of losing one or more players to injury and the identification of the factors necessary to improve team performance (i.e., rebounds, assists, scoring, etc.). Thousands of these simulations can be run to model a wide range of variable changes.

2.3 Other Sporting Simulations

Other sports can benefit from using simulated data as well. In Yacht Racing, a variety of factors on boat design can be tested and winning designs can be built and put into practice (Philpott et al. 2004). In Boxing, an array of both physical and psychological characteristics can be used to determine match winners 81% of the time (Lee 1997).

Hockey game simulation research involves using hidden Markov chains to pattern expected outcomes based upon where the puck is located and the team holding possession (Thomas 2006).

Football games can be simulated using both regressive and autoregressive techniques to determine the factors most responsible for scoring events (Glickman and Stern 1998; Willoughby 1997), as well as Bayesian learning (Stern 1991). Soccer has taken advantage of simulating game play by using Monte-Carlo methods (Koning 2000; Rue and Salvensen 2000).

Data can often hold indications of future performance. By using the right algorithms to identify the key drivers of knowledge, historical data can be used to make accurate predictions.

3 Machine Learning

Aside from statistical prediction, machine learning techniques are another method of providing sport-related predictions. Neural Networks are one of the most predominant machine learning systems in sports. Within neural networks, data sets are learned by the system and hidden trends in the data can be exploited for a competitive or financial advantage. Other machine learning techniques include genetic algorithm, the ID3 decision tree algorithm and a regression-based variant of the Support Vector Machine (SVM) classifier, called Support Vector Regression (SVR).

3.1 Soccer

In a predictive study of Finland's soccer championships, Rotshtein et al. compared the forecasting ability of both genetic algorithms and neural networks (Rotshtein et al. 2005). They first set about classifying the wins into one of five categories: big loss, small loss, draw, small win and big win, where a big loss would be in the range of 3 to 5 point deficit, small loss a 1 to 2 point deficit, etc. From there, they fed past tournament data (e.g., the tournament win/loss performance of each team over the prior 10 years) into both a genetic algorithm and a neural network for training on the most recent 7 years worth of data. The results found that the neural network performed significantly better than the genetic algorithm in all five categories. Overall the neural network had 86.9% accuracy of selecting winners as compared to the genetic algorithm's 79.4% accuracy. The other finding was that the neural network required less time for training compared to the genetic algorithm, which was attempting to optimize the solution set and did not satisfy the study's stopping conditions.

3.2 Greyhound and Thoroughbred Racing

Predictive algorithms have been adopted successfully in non-traditional sports such as Greyhound and Thoroughbred racing.

Predictive algorithms can also be conducted in other non-traditional sports, such as Greyhound and Thoroughbred racing. These types of predictions generally involve machine learning techniques to first train the system on the various data components and second to feed new data and extract predictions from it. While the highlights of several studies are presented here, later chapters will analyze these in more depth.

One machine learning technique that has been used with success in Greyhound racing, is neural networks. In one particular study, a back-propagation neural network (BPNN) was given ten race parameters that were judged by greyhound racing experts as the most important prediction variables. Chen et al. (1994) then evaluated BPNN simulation results in two ways; accuracy of predicting a winner and payout if a bet was placed on the predicted winner. From this work, they found their BPNN to have 20% accuracy and a $124.80 payout as compared to human track experts which managed 18% accuracy and a payout loss of $67.60. Using the same data on the ID3 algorithm, this same group of researchers found better accuracy, 34%, but a lower positive payout, $69.20. It was believed that the BPNN was better equipped to find and capitalize on the races with higher odds, which led to its better payout in spite of lower accuracy.

A follow-up study that built upon Chen's work, examined the influence of predicting Wins, Quiniela (selecting the first two dogs to finish in any order) and Exacta (selecting the first two dogs to finish in order). In this study researchers tested a BPNN with 18 parameters instead of 10, and found similar accuracy results (24.9% Win, 8.8% Quiniela, 6.1% Exacta) but differing payouts ($6.60 loss for Win, $20.30 gain for Quiniela, $114.10 gain for Exacta) (Johansson and Sonstrod 2003). It was suspected that the extra parameters used had a substantial impact on predicting longshot races.

Another follow-up study to Chen's pioneering work, used the SVR machine learning algorithm instead of BPNN. This variant of SVM takes the hyperplane that is used to maximally separate the classes and performs regression estimates against it (Schumaker and Chen 2008). This study was more interested in determining the factors that go into predicting long shots, and would vary its bets from just strong dogs (those that are predicted to finish first) to betting on all dogs (those that will finish above eighth place). From this simulation strategy and a study of all the various exotic wagers, it was found that this system achieved a peak 17.39% accuracy on Superfecta Box wagers (i.e., betting on the first four dogs in any order) when betting on dogs expected to finish between first and second place or better, as compared to random probability at 2.79% (Schumaker 2007).

Predictive measures have also been performed within Thoroughbred Racing. In a study of the factors that lead to racing success, it was found that the motion of a 2-year old thoroughbred's foreleg had a direct relation to its future earnings potential (Seder and Vickery 2005). Horses that were determined by veterinary experts as having good foreleg motion (e.g., a lack of extraneous activity), earned 83% more than those with bad foreleg motion. This has a direct impact on the racing industry as thoroughbred investments can now be screened and improve the chances of selecting a winning steed.

Another important factor in thoroughbred racing is the career length of the horse. Thoroughbreds with longer careers will understandably have better earning potential than those with shorter careers. Using simulation techniques on a pool of potential genetic parents, theoretical offspring can be modeled and their career lengths approximated (Burns et al. 2006). By using these techniques, thoroughbred owners can attempt to maximize the revenue potential of their investments.

3.3 Commercial Products

Commercial systems such as Synergy Online, the Dr. Z System, Front Office Football, and Visual Sports have been developed for sports and gaming predictive modeling.

Aside from research-related simulation and machine learning programs, there are also several programs of the commercial variety. We profile four such systems that are very diverse in their offerings: Synergy Online, the Dr. Z System, Front Office Football, and Visual Sports. As previously mentioned, Synergy Online is one such product that allows individuals access to real-time game data. This data is typically used in console gaming applications to bring a feeling of realism to the gaming experience. The Dr. Z System is of an entirely different type that leverages arbitrage opportunities that arise from odds discrepancies in greyhound, thoroughbred and harness racing tracks. Front Office Football is a simulation product that allows the user to build a professional football team, adjust rosters and make play calls to test optimum team performance. Visual Sports such as Visual Sports Baseball, Visual Sports Golf, and Visual Sports Soccer is a virtualization/simulator where a human participant may engage in an aspect of the sporting activity and the consequence of their actions is simulated on a display screen. This type of technology can instantly analyze the player's movement and suggest changes to improve performance.

3.3.1 Synergy Online

As mentioned earlier, Synergy Online is the driving force behind gaming console products such as NBA Live 09 and NBA Live 10, by providing up to the minute basketball game data to consumers. Their signature gaming product, Digital DNA, models nearly 1,000 characteristics and tendencies of a player (Colston 2009) in order to better model their expected performance and bring realism to the gaming experience (Arnovitz 2009). This product has been implemented in several EA Sports games including NBA Live and NCAA Football.

Synergy Online is also dedicated to collecting basketball-based multimedia and contains an index of live video broadcasts as searchable media. Self-billed as the most visited basketball video repository, Synergy can collect live basketball video, tag the video with an ontology-based form of basketball tags which identify actions within the video, update statistics as events occur and push appropriate content to NBA consumers, such as game footage of their players and the teams they will be facing. With this system, coaches, players and fans can query plays in real-time and receive constantly updating player statistics. Synergy Online also allows live broadcasts to stream to desktops or mobile devices.

From this data-driven approach to parsing broadcast media, it would not be a far stretch to apply an overlay program to query the data further in order to identify pattern trends and returns appropriate video material to back up its conclusions.

3.3.2 The Dr. Z System

The Dr. Z System, named after Dr. Ziemba, a mathematics professor credited with its invention, is a collection of rules and methods based on statistics and empirical research, that allows bettors to take advantage of specific instances in parimutuel betting. At the heart of the Dr. Z System is the idea of the wisdom of crowds to accurately select winning thoroughbreds. From this, odds for each thoroughbred are made with those thoroughbreds with lower odds being expected to finish the race earlier than those with higher odds. However, because humans are involved in this betting process, psychological factors come into play which can override reason and logic. These factors become especially apparent with higher odds horses where the bettor may decide to make a long-shot bet. One reason could be that the bettor has lost a significant amount in wagering and is attempting to recoup those losses quickly through a long-shot bet, that if successful could offset the losses. Another reason could be that the bettor knows of someone who made a killing on a long-shot bet and from that, they may rationalize the odds in such a way that they may believe their chances of winning to be better than the odds would indicate. This leads to the long-shot bias where long-shots are suddenly more favored than they should be and as a consequence the favored horses are shunned. In order for the racetrack to attract an equal amount of wagers, they adjust the odds of the thoroughbreds so that thoroughbreds with a low amount of wagering are given higher odds and a higher payout amount. What the Dr. Z System does, is it allows bettors to recognize conditions of the long-shot bias and calculates the maximum bet to place on a favored thoroughbred with higher than usual odds, based on the expected returns for either a place wager, betting that the horse will finish in either first or second place, or a show wager, betting that the horse will finish in first, second or third place. This system essentially takes advantage of these arbitrage opportunities that come about from psychological means. Not every race will meet the conditions necessary for a statistical arbitrage event. However, the technique, while rooted in thoroughbred racing is generalizable to other forms of parimutuel wagering such as harness racing and greyhound racing. This system was found to be quite effective at identifying and exploiting statistical arbitrage opportunities. Throughout the 1980s and early 1990s this method was used to capitalize on market inequities in the betting pools. However, as the technique drew a more widespread following, the arbitrage opportunities began to disappear. As more bettors were placing Dr. Z type bets, the odds would readjust leading to lower payouts or even losses. This system by itself eventually became a victim of its own success.

While the Dr. Z System may by itself not fall into simulation or predictive modeling, this important technique became a cornerstone in several predictive modeling programs such as HTR Softwares' NeuNet Pro 2.3 (http://cormactech.com/neunet/horses.html) and FindMySofts' Horse Racing Predictor 1.3 (http://horse-racing-predictor.findmysoft.com), which both employ neural networks to learn to find winning horses based on multiple variables including the Dr. Z System.

3.3.3 Front Office Football

Front Office Football, from Solecismic Software (http://www.solecismic.com/index.php), is a professional football simulation program that allows a user to experience what it would be like to be the general manager of a professional football franchise. Users can perform all of the general manager duties such as instituting trades, drafting new talent, managing contracts, negotiations as well as coaching responsibilities such as selecting starting players, calling plays, managing the roster, etc. With each player modeled on real-life data, hypotheses can be created and tested within this simulated environment. A reasonable outcome may carry over into the real world. Along with professional simulation, Solecismic Software also has a collegiate football simulator called The College Years, which allows much the same functionality as Front Office Football, except at the collegiate level. Users of The College Years can inspect and draft high school students based on their data as well as attempt to improve the colleges' standing. Data is interchangeable between the two products, allowing players developed under The College Years to be drafted under Front Office Football. The caveat with this system and all other prediction systems is how far into the future a simulation can go before its results entirely diverge from reality and the errors compound themselves in a manner that ceases to be manageable.

3.3.4 Visual Sports

Visual Sports are live action simulators where a human participant is monitored from a computer system and their actions dictate how the simulation will progress. Also known as indoor live action simulators, these devices typically have a small footprint, a small area for the human participant to stand, a display screen in front of the individual and cameras trained on the participant. In Visual Sports Golf, the system will analyze the golf swing direction, speed and motion in order to simulate ball trajectory on a simulated course. From this usage, a golfer can learn to identify different adjustments in swing that can impact their golfing performance. In Visual Sports Soccer, players can practice making goal shots against a computer generated goalie of increasing difficulty. In Visual Sports Baseball, batting swings can be analyzed and from that the baseball trajectory can be computed, allowing batters to practice their swings and timing against a different variety of pitches and speeds. Visual Sports Basketball allows the participant to practice making shots and Visual Sports Hockey allows a user to practice making shots against a computer-generated goal-tender. All of these simulation technologies allow users to practice their sport, provide immediate and realistic responses to participant actions, and make available the participants' data in reports for further analysis by the participant or interested party.

4 Conclusions

Simulations and machine learning systems are important in the development of modern sports. The ability to apply statistics and rigorous mathematical models to provide instantaneous results is quickly becoming an invaluable commodity. Systems of this type vary from simulations that model an entire upcoming season's worth of data to identify the best chance of winning, to simulations that identify weaknesses in motion and offer advice for correcting them. These simulative systems offer immediate results if their suggested actions are corrected. In this way, the participant can see the fruit of their labor in real-time and having a feeling of accomplishment while they are improving their play-making skills. While still in their infancy, it will be interesting to see what these systems will evolve into in the coming decades.

5 Questions for Discussion

1. Simulations are often seen in a wide arena of sports, but cannot be applied to all of them. What sports would be difficult to simulate and why?
2. What characteristics of a sport are necessary for simulation to be successful?
3. If simulation is not possible on a particular sport, could machine learning work instead? What limitations exist for this technique?

Chapter 7
Multimedia and Video Analysis for Sports

Chapter Overview

Sports information and footage are quickly becoming increasingly available in digital form. Using many of the tools previously outfitted for textual searching, video and multimedia searching and retrieval is becoming more commonplace in sports. Automated methods to watch and listen to games are being used to parse video and render it in searchable form.

1 Introduction

> One company that is leading the way in Major League Baseball's video content is Advanced Media, which has become such a lucrative success that substantial revenues originating from it are distributed to all 30 MLB teams.

Traditional sports statistics are quickly becoming dwarfed by advances in multimedia technologies for sports. Within the past several years, the use of video capture to identify and isolate particular events for subsequent analysis has become much more mainstream. As it stands now, baseball players can visit their teams media room and study a certain pitcher's delivery or their own, to prepare before, or to make adjustments during the game (Lewis 2003). One company that is leading the way in baseball's video content is Advanced Media (Ortiz 2007). Advanced Media handles the digital media content of Major League Baseball, which includes streaming live video to fans and the web-based MLB Game Day tool that portrays an abstraction of the actual game with only the basic game information minus the video. MLB's Advanced Media has become such a lucrative success that substantial revenues originating from it are distributed to all 30 teams.

R.P. Schumaker et al., *Sports Data Mining*, Integrated Series in Information Systems 26, DOI 10.1007/978-1-4419-6730-5_7,

Baseball is not the only sport to benefit from multimedia technologies. Basketball players can use similar searchable video footage services to find all video relating to particular types or locations of shots or defensive moves (Sandoval 2006). Before this technology became available, teams would often have to wait several days to receive game footage and then parse it according to their own needs, but now footage is almost instantaneously streamed to players, coaches, and scouts alike. It used to be a tedious process to identify and retrieve particular video sequences, but now making queries such as in soccer, "find corner kicks resulting in a goal in the final 2 min" will parse the natural language query, search for tagged video sequences that best match the query's constraints, and return the appropriate video footage for further analysis.

2 Searchable Video

The process of automated video retrieval may seem to be a daunting task. While broadcast video may be considered a wide domain with many extraneous frames that do not contribute to the game, such as fan shots, city/stadium shots, sideline shots, announcers or cutesy graphics/ads cluttering up a portion of the video; sports events can be recognized and sequenced as a particular order of actions (Roach et al. 2002). This process is relatively straight-forward because sports events can be thought of as a well-defined sequence of events that are conditionally-based on game-based actions. The structure of the game can be harnessed and much more easily partitioned into sequences of video action. As an example, to find a particular baseball player's at-bat sequence, many facts about the game are known in advance, such as the batting order. Similarly, when transitioning from one batter to the next, the first batter will either get on-base or head for the dugout, both of which can be identified as separate events by a multimedia system and tagged as such. Then the process of retrieving specific video sequences becomes a matter of querying the tagged material.

Tagging sports events in real-time can be done manually by domain experts or automatically by identifying a sequence of events (e.g., a basketball that approaches the rim of the hoop can be tagged as a shot). Automatic tagging typically takes advantage of changes in the video, such as pans, zooms, fades and cuts which signals that a new video segment has begun or an event is taking place (Truong and Dorai 2000).

Tags can be metadata or a simple descriptor of events. Consider the following metadata tagging using XML (Babaguchi et al. 2007):

```
<AudioVisualSegment>
 <StructuralUnit>
      <Name>at-bat</Name>
 </StructuralUnit>
 <TextAnnotation>
    <FreeTextAnnotation>Arias</FreeTextAnnotation>
    <StructuredAnnotation>
   <Who>
      <Name>Arias</Name>
      <Name>Kudou</Name>
   </Who>
    </StructuredAnnotation>
    <KeywordAnnotation>
   <Keyword>SoloHomeRun</Keyword>
   <Keyword>OpenTheScoring</Keyword>
    </KeywordAnnotation>
 </TextAnnotation>
</AudioVisualSequence>
```

In this sequence, one can easily understand that the event is a solo homerun by Arias, which led to the team's initial score of the game.

Many new sports video searching and analysis tools have been developed recently, including: SoccerQ, blinkx, Clipta, SportsVHL, Truveo, and Bluefin Lab.

2.1 SoccerQ

One notable multimedia tool is SoccerQ, which allows users to store, manage and retrieve soccer game video sequences (Chen et al. 2005). This program supports basic queries such as: select video/shot/variable from search_space [where condition]. The "variable" term can refer to a particular team, where "search_space" can be limited to certain sub-categories, such as men's or women's soccer, etc. As an example, a query may be "select all corner kick shots from all female soccer videos where the corner kick resulted in a goal event occurring in 2 min," as shown in Fig. 7.1.

From Fig. 7.1, the left side of the SoccerQ application has only "female soccer" selected to limit the search space. The right side shows two events selected, "Corner Kick" and "Goal." To the far right, the temporal precedence of Event A (Corner Kick) starts Event B (Goal) is selected. At the right center, 2 min is selected. The far bottom shows the four scenes that match the criteria. While the SoccerQ application constrains user input to preset values, other areas of multimedia retrieval are focusing on using natural language queries, similar to how a web-based search engine would approach the problem.

Fig. 7.1 SoccerQ video retrieval (Chen et al. 2005)

2.2 Blinkx

Blinkx.com touts itself as the world's largest and most advanced video search engine that deals primarily with sport-related video (blinkx.com 2009). This video content provider has partnered with FoxSports and receives approximately 4 h of sports video per day which is made available to users. Aside from keyword tagging of video sequences, blinkx also analyzes the closed captioning text to gain information about game events as well as machine identification of game sequences. In addition to their collection of FoxSports media, blinkx also collects media from other sources and currently advertises a video collection size of 35 million hours.

2.3 Clipta

Clipta.com is another sports source for video products. This site focuses on the retrieval of specific video and makes other video-based recommendations according to the browsing search habits of the individual and how well they match up to prior searching behavior (clipta.com 2009). Their algorithm, the V-Rate system, attempts to re-rank video results depending on the user's perceived searching tendencies in order to provide a better video retrieval experience.

2.4 SportsVHL

SportsVHL is a novel web-based product that does not focus exclusively on professional sport video, but also maintains video on college and high school sports. Built as a searchable video retrieval engine, Sports VHL can narrow down searches based on sport, geographic area and even age level of participants (SportsVHL 2009). SportsVHL can also be used by coaches and scouts as a recruiting tool with a built-in set of pre-defined queries such as names, sport, position, speed, jump height, height, weight, geographic area, etc. This allows coaches and scouts to see video of potential players that match their criteria without the expense of an extensive scouting department.

2.5 Truveo

Truveo is another video-based searchable retrieval engine, with the twist being that it searches other websites for the video content. Launched in 2004 and bought by AOL in 2006, Truveo boasts the ability to search through over 300 million videos, with 20 million on sports alone, from dozens of websites (truveo.com 2009). The majority of their indexed content resides on sport-related league sites, such as MLSnet.com, MLB.com, NHL.com, NBA.com, etc.

2.6 Bluefin Lab

Bluefin Lab, a spinoff from MIT's Media Lab, is similar to blinkx.com, in that they plan on automatically sequencing video using a variety of techniques from closed captioning, video sequence transitions and game event identification (Moore 2009). Scheduled to debut sometime in 2010, this startup leverages several MIT Media Lab staffers and has a sizeable venture capital investment.

3 Motion Analysis

Motion analysis in sports research is generally concerned with object tracking and trajectory. Baseball is a hotbed of motion analysis research where not only pitching mechanics and ball trajectory are analyzed, but also the motion of batters as they approach different pitches. Object tracking and trajectory analysis usually starts with a baseline video image where the item of interest is selected or some reference point of known size is recognized (e.g., a human can be of an estimated size for comparison to the size of a ball) (Chang and Lee 1997). Once the system has been calibrated, it can track the motion of the intended target.

Another method in motion analysis is to break videos into three component parts: the background or camera motion, the foreground object motion, and shot or scene changes (i.e., a different camera view) as a result of an external edit (Roach et al. 2001). From this approach, videos can be segmented and foreground motion tracked.

One of the techniques to identify trajectory is to filter the video frames. Using baseball as an example, the video would be filtered to identify all the white objects in the frames, white being the color of the ball. From there, eliminate any of the white elements that do not conform to the size or shape of the baseball. Now that a candidate list of balls has been identified, the process is repeated on the following frames and the white speck that exhibits motion between the pitchers mound and the batters box is the baseball (Chu et al. 2006). Once the trajectory has been identified, the type of pitch can be extracted as well. Curve balls typically exhibit a high arching motion which stands in contrast to a slider with lateral or sharp downward motion. These systems have fairly high accuracy, around 90% recognition (Chen et al. 2007). Aside from baseball, these techniques can be used on other sports as well such as soccer where it is used to track ball location and speed, player location, shot distance and the distance a player has run (Bialik 2007; TRACAB 2007; Yow et al. 1995), in football (Ding and Fan 2007) and even tennis (Takagi et al. 2003).

4 Conclusions

The field of searchable sports video is evolving. Only 2 years prior to this writing, these tools were not widely available and required a lot of manual labor to achieve today's automated results. The tools that link multimedia retrieval to data mining are promising and are quickly being developed. As the technology becomes more mature and questions of retrieval research begin to be answered, this emerging market shows a lot of opportunities over the next several years.

5 Questions for Discussion

1. What characteristics of a sport would make it easier for video indexing and searching?
2. What sport or sports would be difficult to use automated video parsing techniques and why?
3. Can motion analysis of video be applied to other sports and how?

Chapter 8
Web Sports Data Extraction and Visualization

Chapter Overview

How is it that we value data? Is a simple repository of data all that we need? It used to be that carrying a copy of Total Baseball was all that was ever needed, as it provided a historical perspective of player data that was adequate for our needs only a decade ago. Then as sabermetrics began to awaken the sporting world's desire for more data and consequently new ways of analyzing that data, data itself began to evolve. Data first moved from static pages of written form to online resources. While this step was simply a change of venue, data was still data, but it soon began to become more. Web applications began to sort this data into leaderboards on a whole host of different statistics, thus entered information. From there, the applications evolved further, exploring the graphical realms of presentation, pushing that information into knowledge. It is amazing to think how quaint our memories of carrying a printed copy of *Total Baseball* are by today's standards.

1 Introduction

Data is inextricably linked to sports performance. The more available the data, the better able we are to measure and compare performances. Online data sources have become more abundant as the proliferation of the Internet has increased along with the demand for instant, accurate and easy to use tools. These data sources range in character from officially sanctioned league repositories to multi-sport third parties. Besides the data itself, the use of this data has begun to take novel and unexpected turns. Such as a security tool used at several college campuses to track fan violence and respond to incidents quicker and with the appropriate amount of force. Data, its application, and the questions we need answered are all constantly evolving.

R.P. Schumaker et al., *Sports Data Mining*, Integrated Series in Information Systems 26, DOI 10.1007/978-1-4419-6730-5_8, © Springer Science+Business Media, LLC 2010

2 Web Data Sources

Many new sports web data sources have been developed by their respective sports league's official governing body in recent years, such as: MLB.com, NBA.com, NFL.com, and NHL.com. Most of these sites aggregate the raw data into easy-to-understand charts, graphs, and projections for the general public.

There are many more web data sources than there were even a few years ago. Many of these data sources originate from their respective sports league's official governing body, however, there are also a healthy amount of third-party sources that offer data as well. While some of these websites are subscription-based models, others offer their data for free and make their money from web advertising and other revenue sources. In either model, the proliferation of data over the past decade that has been made easily accessible is staggering. Even better, many of these sites aggregate the raw data into easy-to-understand charts, graphs and tendency projections, where users can drill-down through the aggregated data to discover new patterns and view the data that goes into the aggregated value. Oftentimes, this style of reporting allows easy access to determine causes behind outlier and exception data.

2.1 Baseball

Baseball has many different web data sources. From MLB.com, Major League Baseball's governing body which provides pitching and batting tendencies as well as gameday abstractions of real games, to Retrosheet.org which provides historical game scoring and a recording of relevant game events.

2.1.1 MLB.com

MLB.com, the governing body of Major League Baseball, contains a wealth of sortable data and a variety of colorful and easy to understand graphical depictions of player performance. For instance, users can query the system to find statistics based on batter-pitcher matchups. As an example, Cleveland's Shin-Soo Choo had a 0.405 batting average against Kansas City pitching in Kaufmann Stadium over his 37 at-bats during the 2009 season. By knowing how well players are performing in particular environments and against particular teams, managers can reap the reward of playing the statistics and hope for a significant payoff.

While these baseline statistics are free but somewhat limited, MLB.com has also instituted premium subscription-based content through their MLB Gameday service.

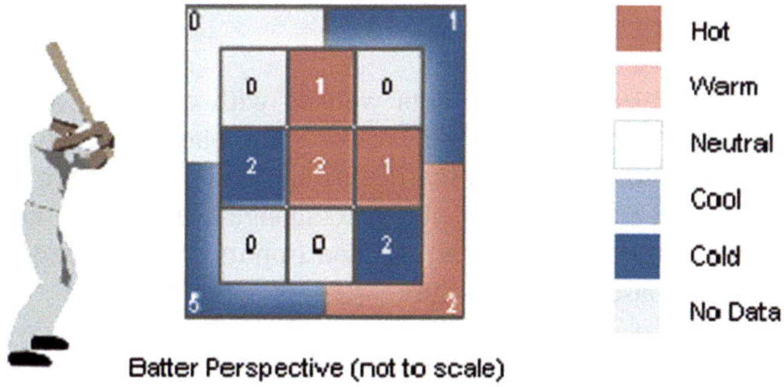

Fig. 8.1 Hot/cold zones, courtesy of http://www.mlb.com/mlb/gameday

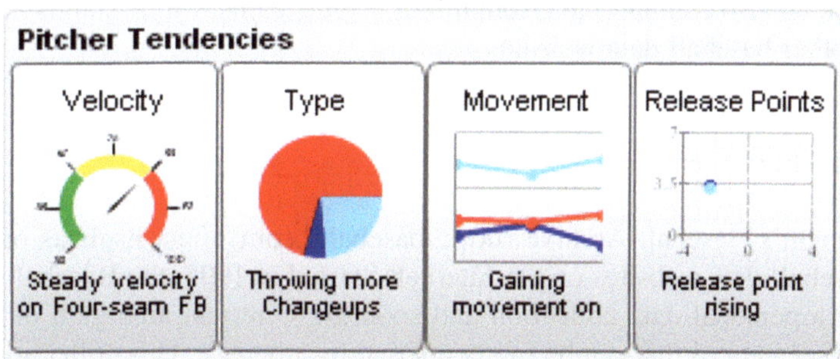

Fig. 8.2 Pitching tendencies, courtesy of http://www.mlb.com/mlb/gameday

Users of this service can see advanced graphical descriptions of gametime data such as a batter's Hot/Cold zones as shown in Fig. 8.1. In this graphic, the strikezone is divided into nine equal areas. The color of the area denotes the performance of the batter in that part of the strikezone. Red is hot, meaning that the batter has a high percentage of success hitting pitches in that area versus dark blue or cold, meaning that the batter has a difficult time connecting with pitches in that area. Aside from color coding, numeric data is provided in each area indicating the number of pitches thrown in that particular area.

Another MLB.com Gameday tool is the Pitcher/Batter tendencies chart which evaluates performance throughout the game. In the case of pitching, the Pitcher Tendencies tool, shown in Fig. 8.2, provides pitch velocity, type of pitch, what movement the pitched ball is exhibiting and release points, the position relative to the pitcher's body where the ball is released from the pitcher's hand. Over the course of the game, this data can indicate pitching problems, such as fatigue – the velocity steadily decreases, a loss of movement on the different types of pitches, an over-reliance on certain types of pitches in later innings, and changes in release points which can impact ball placement in the strike zone.

2.1.2 Retrosheet.org

Retrosheet.org is a historical game data website with complete and continuous boxscore data since 1952, textual narratives of game play for nearly every major league game of record, player transaction data, standings, umpire information, coaching records, and ejections of players and managers alike. This data collection can be downloaded via formatted text files and imported into spreadsheets, document editors, or databases for ease of use.

2.1.3 Baseball-reference.com

Baseball-reference.com is another baseball statistics source that holds historical and current data, awards, league information, and a blogging feature where users can share information and insights. Aside from Major League player data, baseball-reference.com also contains data from minor league teams, which stands in contrast to many other baseball data websites.

2.1.4 Baseball Archive

Sean Lehman's Baseball Archive (http://baseball1.com) touts itself as one of the oldest baseball data websites on the Internet. Started in 1995, the Baseball Archive started as a personal data collection and soon grew into an amalgam of multiple baseball data sources that can be freely queried by any user. This collection was an answer to Bill James' complaint during the 1980s that baseball statistics were not freely available to the public.

2.2 Basketball

Much like baseball, basketball also has a wealth of online sources of data. From NBA.com which provides historical game data, game-based charts and other reference material, to Basketball-reference.com with their unique player matchups and other insightful analysis.

2.2.1 NBA.com

The governing body of professional basketball has an extensive array of data available to users. This data ranges from basic statistical rankings by both player and teams, to more sophisticated plus/minus ratings and interactive graphics of player point shooting. As an example of their statistical coverage, NBA.com displays daily leaders in categories such as points, rebounds, assists, steals, blocks, and three pointers. Sortable statistics based on player or teams, can also provide useful insights into player performance as well as plus/minus statistics which identify the five player combinations that score the most points while holding their opponents to the least amount of points. Figure 8.3 demonstrates the top five player combinations using the plus/minus rating. Note that Mavericks players Kidd, Dampier, Nowitzki, Marion, and Terry have so far during the 2009–2010 season scored 172 points while holding their opponents to 105 points. This leaves this combination with a plus/minus rating of +67.

The graphical description of data is by far the most interesting aspect of NBA. com. These graphical descriptions range greatly in content and interactivity. However, they all share ease of use and the ability to communicate information to the user quickly and intuitively. As an example, each completed NBA game provides textual boxscore data relaying the events of the game, a news article describing the game in depth and a graphical *Stats at a Glance* which conveys the important game data elements of field goal percentage, three pointers and free throw percentage – see Fig. 8.4. As shown in this figure, it would appear that the Hawks dominated the Knicks in both field goal percentage (54.3 to 47.1%, respectively) and three pointers (46.2 to 20.8%, respectively). However the Knicks managed a better free throw percentage (88.9 to 83.3%, respectively).

Aside from the intuitive nature of *Stats at a Glance*, NBA.com also provides an interactive web application called Courtside Live – see Fig. 8.5. Within this environment, the user is provided with the basic boxscore data across the top and left-side of the application and in the center is a depiction of the basketball court showing all the shot attempts by both teams, coded by color; red is the Hawks and orange is the Knicks. A circle indicates that the shot attempt was successful whereas the x means an unsuccessful attempt. These attempts, circles and x's, are interactive and a user can mouse over them to reveal additional details in a pop-up box, as shown in the bottom center. In this pop-up, we can see that Al Harrington attempted a shot from this location, right-side beyond the three point line, during the third quarter with 4:01 remaining and missed the basket. Furthermore, we could look at other shots by the Knicks and quickly see that Harrington attempted many three pointers and missed all of them. A graphical tool such as this can quickly identify favored areas that players like to take shots as well as areas where player are and are not successful at those shot attempts.

Top Five-Player Combinations

PLAYER 1	PLAYER 2	PLAYER 3	PLAYER 4	PLAYER 5	TEAM	+	-	+/-
J. Kidd	E. Dampier	D. Nowitzki	S. Marion	J. Terry	Mavericks	172	-105	67
M. Bibby	J. Crawford	J. Johnson	Jo. Smith	A. Horford	Hawks	155	-100	55
J. O'Neal	Q. Richardson	D. Wade	M. Beasley	M. Chalmers	Heat	293	-245	48
K. Garnett	R. Allen	P. Pierce	K. Perkins	R. Rondo	Celtics	490	-443	47
M. Camby	B. Davis	R. Butler	C. Kaman	E. Gordon	Clippers	178	-137	41

Fig. 8.3 Top five plus/minus rankings, courtesy of http://www.nba.com/statistics/plusminus/plusminus_sort.jsp?pcomb=5&season=22009&split=9&team=

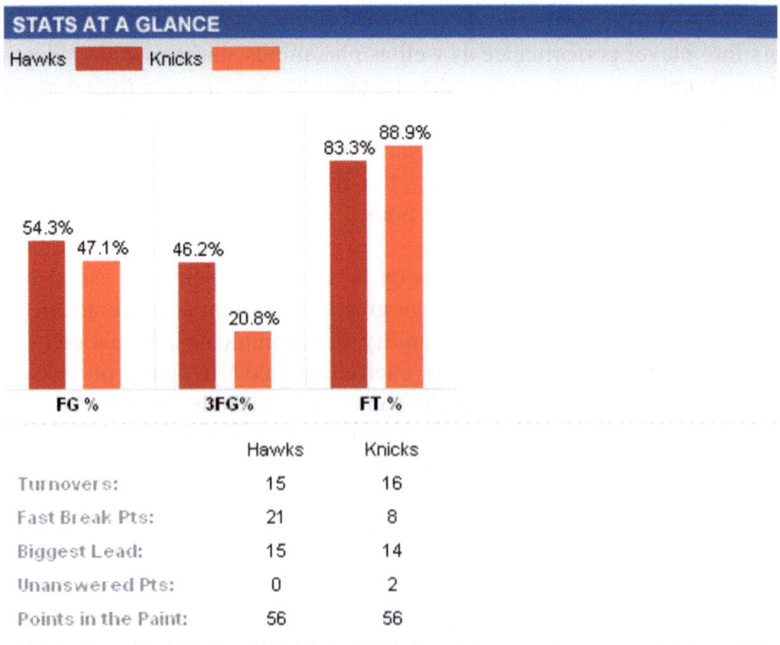

Fig. 8.4 Stats at a glance, courtesy of http://www.nba.com/games/20091111/ATLNYK/gameinfo.html

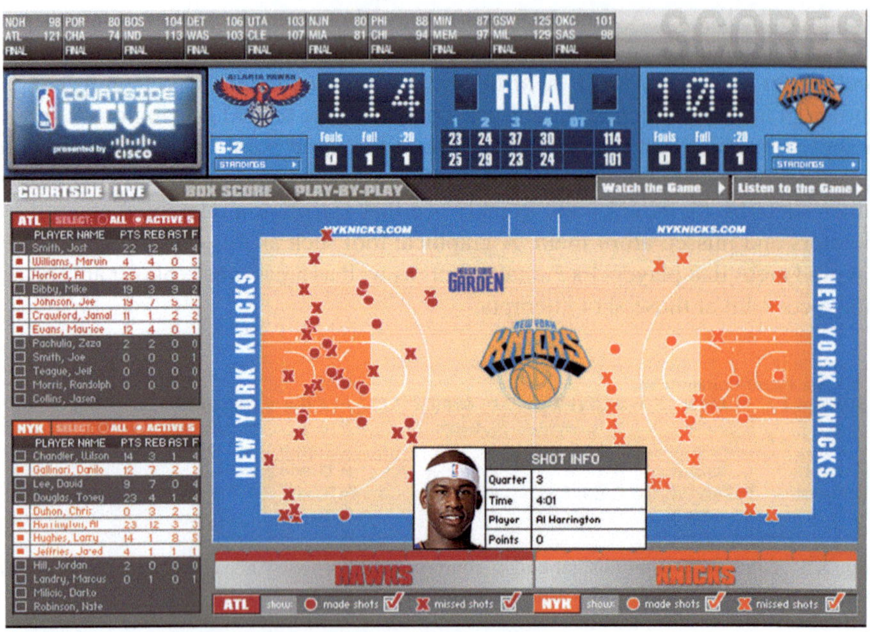

Fig. 8.5 NBA courtside live courtesy of http://www.nba.com/csl/index.html?gamecode=20091111/ATLNYK

2.2.2 Basketball-reference.com

Basketball-reference.com was created in 2003 and is similar in goals to Sean Forman's baseball-reference.com. This site attempts to be comprehensive, well-organized, and responsive to data requests. The basketball data is relatively straight-forward and easy to navigate.

2.3 *Cricket*

An amazing amount of cricket statistics has been collected and made available online in recent years. What was once housed only in the Wisden Almanack or kept locked away by Cricket score keepers, is now easily accessible to Cricket enthusiasts. Websites such as CricInfo.com and Howstat.com both provide data on test and ODI (One Day International, a form of cricket) matches for historical or real-time needs.

2.3.1 Cricinfo.com

ESPN's cricinfo.com bills itself as the top cricket website that includes cricket news, analysis, historical data as well as real-time matchups. This website includes the StatsGuru tool, which is a sortable stats tool that allows users to drill through the data to find interesting nuggets. As an example, Fig. 8.6 demonstrates the top players in test matches between India and Pakistan from 1978 to 2009, rank-ordered

View overall figures [change view]
Primary team India
Opposition team Pakistan
Ordered by runs scored (descending)

Page **1** of **3** Showing **1 - 50** of **111** First Previous Next

Overall figures

Player	Span	Mat	Inns	NO	Runs
SM Gavaskar	1978-1987	24	41	4	**2089**
DB Vengsarkar	1978-1987	22	35	6	**1284**
V Sehwag	2004-2006	9	14	0	**1276**
R Dravid	1999-2007	15	26	3	**1236**
M Amarnath	1978-1987	18	28	4	**1080**

Fig. 8.6 Top runs scored stat between India and Pakistan test matches, courtesy of http://stats.cricinfo.com/ci/engine/stats/index.html?class=1;opposition=7;team=6;template=results;type=batting

Batting

Ground	M	Inns	NO	50s	100s	HS	Runs	Avg
Barabati Stadium	1	1	0	0	0	5	5	5.00
Brabourne Stadium	1	2	0	1	0	67	71	35.50
Chidambaram Stadium	12	21	4	3	3	*236	1018	59.88
Chinnaswamy Stadium	8	12	1	3	2	172	600	54.55
Eden Gardens	8	15	2	1	2	*182	583	44.85
Feroz Shah Kotla	9	14	0	2	3	121	668	47.71
Gandhi Stadium	1	1	0	0	0	5	5	5.00
Green Park	9	14	0	5	1	176	629	44.93
Sardar Patel Stadium	2	3	0	2	0	90	154	51.33
Sawai Mansingh Stad	1	2	0	0	0	24	24	12.00
Vidarbha Cricket Gr	2	3	0	3	0	74	188	62.67
Wankhede Stadium	11	20	0	3	5	205	1122	56.10
Overall (12)	**65**	**108**	**7**	**23**	**16**	***236**	**5067**	**50.17**

Fig. 8.7 Gavaskar's batting statistics by Indian stadium, courtesy of http://howstat.com/cricket/Statistics/Players/PlayerCountries.asp?PlayerID=0595

by runs scored. As shown in this figure, Sunil Gavaskar is top on this list, scoring 2,089 runs over 24 test matches between 1978 and 1987.

2.3.2 Howstat.com

Howstat.com is another Cricket data repository with many features. Aside from having historical and real-time data, howstat.com also contains a superb searching and sorting application to make data requests simple and easy to use. If we were to expand upon our knowledge of Gavaskar and analyze his performance in test matches by Indian stadium, we would produce Fig. 8.7. Note that his highest average (62.67) occurred within Vidarbha Cricket Ground. Furthermore, these statistics can be drilled down even father to reveal that 74 of those runs occurred on Dec. 27, 1986 in the first inning.

2.4 Football

American football also has its share of data and statistics. While not as rich as baseball and football because of the shorter season length in respect to games played; football has its share of interesting glimpses into performance and non-conventional areas such as home field advantage and recommended strategies for fourth down situations.

2.4.1 NFL.com

The National Football League, governing body of American football, also keeps data on their official league website of NFL.com. This data is fairly standard, composed of top ranked players, player comparisons, and team statistics. This site also has Game

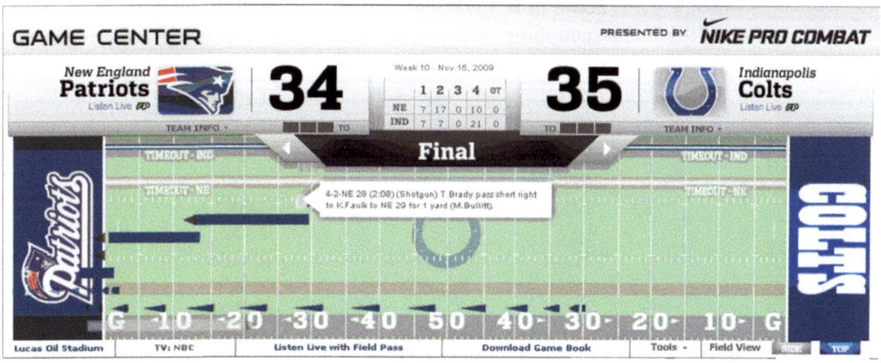

Fig. 8.8 NFL game center, courtesy of http://www.nfl.com/gamecenter/2009111512/2009/REG10/patriots@colts

Center, which is an interactive graphical description of the game, allowing users to search for and find plays within the game. Figure 8.8 shows the Game Center graphic of the Colts – New England game and the box in the center describes in detail the highlighted play where Brady attempted to go for it on fourth down, failed, and Manning and the Colts drove the ball the other way for the game winning touchdown.

2.4.2 Pro-football-reference.com

Built in the same style as baseball-reference.com, pro-football-reference.com provides ample statistics, analysis and commentary to hold any football enthusiasts interest. Users can peruse reams of data regarding coaches, the draft, historical boxscores and team rosters over the years, much of which is unavailable through the official league's website. Pro-football-reference approaches data and knowledge from the sabermetric point of view, asking questions and then seeking the answers. These types of interactions are evident within the website's blog application. As an example, one user was interested in the notion of home field advantage for teams with new stadiums. This user set about gathering win-loss data for teams based on how many visits they made to a new stadium, see Table 8.1. The results were interesting, if a visiting team was playing in a new stadium and it was their first visit there, they had a 35.7% chance of winning. Contrast that with the winning percentage of a team that has made many visits and their winning percentage increases dramatically.

2.4.3 AdvancedNFLStats.com

AdvancedNFLStats.com is a more research-driven collection of football enthusiasts that share their insights and passion for the sport. While this website does not

Table 8.1 NFL home field advantage, courtesy of http://www.pro-football-reference.com/blog/?cat=52

Visit	Wins	Losses	Win percentage
1	134.5	242.5	0.357
2	60	113	0.347
3	50	54	0.481
4	29	34	0.460
5	20	30	0.400
6	19	25	0.432
7	21	18	0.538
8	6	10	0.375
9	7	7	0.500
10	4	5	0.444
11	3	2	0.600
12	2	1	0.667

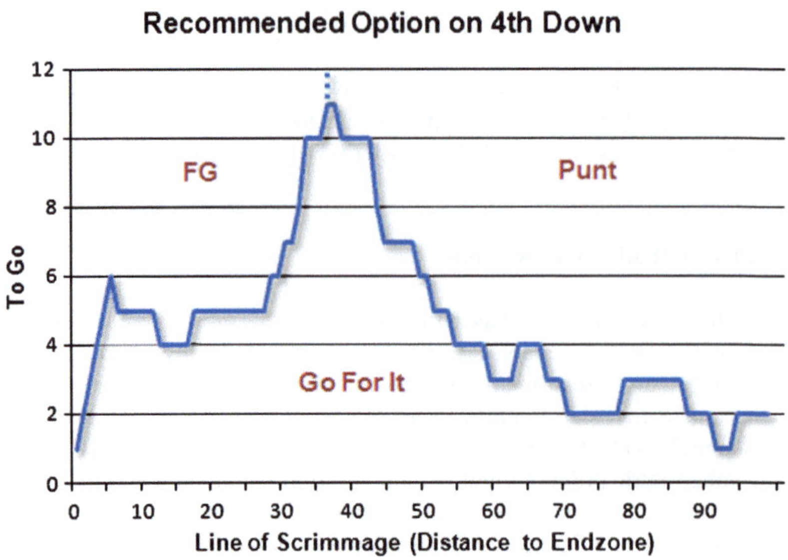

Fig. 8.9 Recommended play calling for fourth down situations, courtesy of http://www.advancednflstats.com/2009/09/4th-down-study-part-4.html

contain the usual fare of historical or real-time data, it instead focuses on sabermetric-styled creations such as game excitement rating, comeback expectancy, etc. As a part of their research focus, one study analyzed recommended plays based on fourth down, the distance remaining in the down and the distance to the endzone (see Fig. 8.9). Most NFL teams will punt or attempt a field goal in fourth down situations as a way of mitigating risk and playing it safe. As shown through the research, teams should attempt to *go for it*, if they have less than 4 yards remaining and are approximately 30 yards away from the endzone. As the distance from the endzone increases, the play recommendations will change as well.

2.5 *Hockey*

Hockey has experienced a data-centric rebirth in recent years. In only a few short years, hockey data has transitioned itself from proprietary and housed mostly in books, to being made freely available online. Data found for professional hockey now rivals that of the other major sports.

2.5.1 NHL.com

Professional hockey's governing body, the National Hockey League, also presents data and statistics to users through their website of nhl.com. This website is as rich in material as both baseball and basketball's official websites; it offers the standard complement of leaderboards, team statistics and historical game comparisons. It is more involved in its presentation of statistics and provides more depth with the quantity of sortable statistics used and the ability to drill down through the data. Similar to basketball, nhl.com also has interactive graphics that can allow users to graphically query the data such as shots, goals, penalties and fights, as shown in Fig. 8.10.

2.5.2 Hockey-reference.com

Hockey-reference.com is a part of the baseball-reference.com, basketball-reference.com and pro-football-reference.com family. This website offers all of the

Fig. 8.10 Rangers – Capitals GameCenter, courtesy of http://www.nhl.com/ice/gamecenter.htm?id=2009020289

conventional sortable historical and real-time hockey data based on players, teams, leagues, coaches, etc. A blog application allows users to share ideas and insights with one another.

2.6 *Soccer*

Similar to hockey, soccer has also experienced a rapid online data presence in recent years. While data on European and World Cup countries is still evolving, American soccer data has launched itself to the forefront of data accessibility and unique ways of presentation. MLSnet.com's GameNav system is one such innovative product that provides video footage abstractions of game events, in a searchable timeline.

2.6.1 MLSnet.com

MLSnet.com, American major league soccer's governing body, has made game data available in fascinating ways. Aside from the typical repast of historical and real-time player, team, and league data available in sortable form, their GameNav application abstracts video footage of game events into a console game-like environment (see Fig. 8.11).

Fig. 8.11 Galaxy – Dynamo GameNav abstraction of a goal, courtesy of http://matchcentre.stats.com/stats-football.asp?g=2009111307&lg=MLS& domain=mlsnet.stats.com

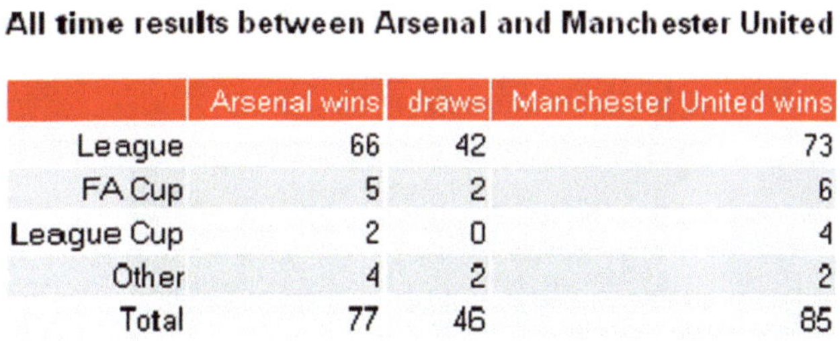
All time results between Arsenal and Manchester United

	Arsenal wins	draws	Manchester United wins
League	66	42	73
FA Cup	5	2	6
League Cup	2	0	4
Other	4	2	2
Total	77	46	85

Fig. 8.12 Arsenal vs. Manchester United all-time results, courtesy of http://www.soccerbase.com/head2.sd?team2id=1724&team1id=142

In this figure, the video footage is from a game between the Houston Dynamo and LA Galaxy and depicts a goal by Berhalter at the 104-min mark. While this figure only relays a still-shot image, the media presented to the user shows life-like movements and one could imagine themselves watching the play from the field level.

2.6.2 Soccerbase.com

For fans of English Premier football, soccerbase.com provides the expected historical data on players, teams and leagues. This data is sortable and its unique twist is the ability to compare players or teams, as shown in Fig. 8.12. In this figure, we can see that in overall play, Manchester United has won against Arsenal, 85 to 77 times, with 46 draws.

2.7 Other Sport Sources

There has also been a rise of multi-sport data repositories. These websites generally track and maintain data for the official sport governing bodies and provide it as a service to subscribing members and the media, as is the case of Stats.com. Other websites instead focus on providing betting information, such as lines and spreads on matches, as is the case of atsdatabase.com.

2.7.1 Stats.com

Stats.com is a multi-sport data repository that contains historical and real-time sport data. This subscription-based service works with official league bodies to provide users, teams and the media with up to the second multimedia in the form of textual score updates and interactive graphics indicating shot progression and scoring. The amount of data they maintain is exhaustive and the graphics provided are top notch. As an example, Fig. 8.13 shows an example of their Cricket dashboard

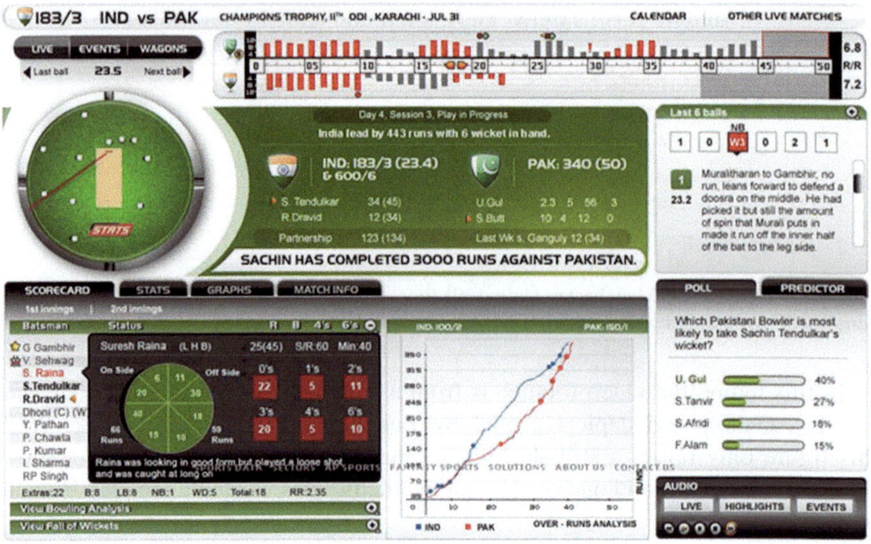

Fig. 8.13 Stats.com Cricket Dashboard, courtesy of http://www.stats.com/cricketdashboard.asp

applicationthat quickly conveys the most important game elements in a quick and easy to use manner.

2.7.2 Atsdatabase.com

Atsdatabase.com is a subscription based service that provides line odds and betting advice to bookmakers and individuals alike. The free data provided on the website is fairly basic in its composition and is geared toward enticing users to subscribe to their paid-services.

3 Extracting Data

While most online data is securely tucked away in data vaults of individual organizations, some sports websites offer their data freely in the hopes of sharing insights and revelations regarding the data usage.

Once the data has been identified and found, extracting it for subsequent analysis becomes of paramount importance. While most online data is securely tucked away in data vaults, only accessible through proprietary web applications, some websites offer their data freely in the hopes of sharing insights and revelations regarding the data usage. Several notable open and closed system programs, take advantage of the presented data in novel and interesting ways, such as GIS mapping of stadium incidents, which is being implemented at many college campuses as a way to control crowd violence as early as possible and make the game safe and enjoyable for fans.

3.1 Programs

There are not many commercial programs available to actively mine data for users for several reasons. One of which is the lack of homogeneity of data across websites. Each website may code their data differently and change schemas from time to time. This makes it difficult for a generalized program to retrieve data versus a specialized one that is constantly updated. Another reason is the load placed on a webserver that is being mined, might be too excessive given their bandwidth considerations. Oftentimes, data repositories that pay by their bandwidth usage, will attempt to block out or ban users that exceed what is considered normal usage, and some data mining programs may fit the role of non-normal usage. As a result, most data extraction programs are written for the sole purpose of mining one particular repository for research purposes and attempt to be careful in their activity so as to not arouse attention and/or a ban from the website.

However, there are other instances where the data and the application are intertwined or are offered by the same organization. Here the rules change, where the mining of data is a needed step. Applications of this nature include mitigating security risks at games and improving multimedia footage acquisition from fast-moving events.

3.1.1 Crowd Control Programs

One novel use of data in a usable application, comes in the form of crowd control programs. The need for these types of systems has slowly gained traction as the rise of sport-related fan violence has escalated. It used to be that security personnel would personally witness a disruptive event, or be alerted by other personnel or fans, and then attempt to take control of the situation themselves. Oftentimes, by the time security has been alerted, the situation has escalated, requiring more personnel to control the violence. One novel way that many college campuses are striking back, is through a text messaging service, where fans can alert security to potential problems, before the problem has a chance to escalate (Lavigne 2009). How it works, is a fan will send a text message via their cell phone to a special number, and provide their seat location and a brief description of the disturbance. The message is then relayed to a centralized security center, where security personnel can pan cameras on the reported section

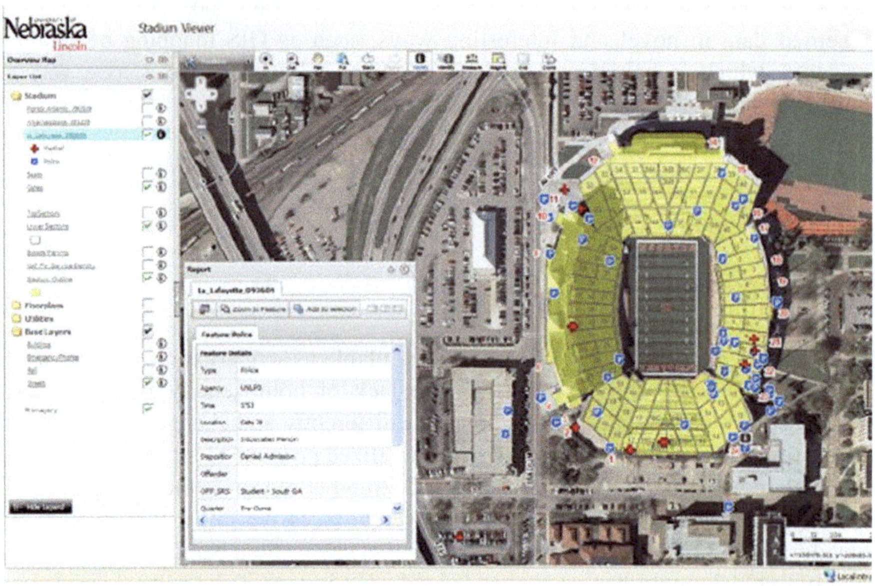

Fig. 8.14 Nebraska's crowd control program using GIS mapping, courtesy of UNL police

and ascertain what level of security response to dispatch. While the system has had its share of false calls and other mischief, these represent only a small fraction of the real messages arriving. Furthermore, some colleges are tabulating this disturbance-related data to determine problem sections and also as a way of allocating security personnel. Through their analysis, they found that most of the disturbances occur in student sections and generally have some form of alcohol involved. Figure 8.14 shows such a mapping of problem areas in the University of Nebraska's football stadium.

3.1.2 Tracking High Speed Sports

Another novel use of sport data, comes from the field of multimedia sports research. High speed sports, skating, bobsledding, skiing and the like, have all had problems with generating clear stop-frame footage of the event. Typically, the speed of the contestant creates a blur on standard broadcast footage equipment, because of the frame rates used. One way to obviate this problem is to move the camera farther away from the event, however, the drawback is less detail. What commonly occurs is that the camera is positioned closely to the action and pans quickly as the players cross the field of vision. The problem arises when computerized tracking tools are monitoring the video footage, tracking a particular contestant's movement. This quick panning motion of the camera coupled with blurring of the action, overwhelms many tracking systems. However, some recent research has found a way to mitigate these problems by applying transformation matrices to each frame and then calculating movement between, while taking advantage of other contextual clues, such as a stationary object (Liu et al. 2009). While it may seem an obscure usage of data, in this case the data comes in the form of multimedia footage. There has been a growing trend of using computers to identify competitive advantages through video. This concept is generalizable enough to be applied to other fast moving sports with quick camera pans where computers are using the feed for targeting movement.

4 Conclusions

Sport-related data has come a long ways over a short period of time. From being locked away by team scorekeepers, to being tabulated in book form by enthusiastic fans, to being published online as raw data, then sortable data and later as refined graphics with queryable support; sport-related data has transitioned much over this past decade. With so much interest in the data from fans, trivia buffs, sabermatricians, fantasy team owners and the like, the ability to provide concise data that is easy to search, easy to sort, and easy to use will be of paramount importance in the coming years. Integrating multimedia into the existing data cavalcade will be the next logical progression. Where we go from there will be up to us.

5 Questions for Discussion

1. Pick a sport and research the data associated with it. How could the treatment of the data within this sport be improved?
2. Describe your vision of sport-related data, 5 years from now. Ten years from now.
3. Many sport-related websites tend to look similar or have the same types of data available. What potential drawbacks exist with this type of approach?

Chapter 9
Open Source Data Mining Tools for Sports

Chapter Overview

Open source development has become more prominent in recent years in a multitude of software areas. In the domain of data mining tools, several solutions have gained significant acceptance such as Weka and RapidMiner. Both tools share the same underlying learning algorithms, however, their approach to displaying results, are very much different.

1 Introduction

Both Weka and RapidMiner are excellent open source tools for sports data mining.

Both Weka and RapidMiner are excellent open source tools that can leverage multiple algorithms, allowing users to rapidly explore and anlyze their sports data however they see fit. This means that users can run their data through one of the built-in algorithms, see what results come out, and then run it through a different algorithm to see if anything different stands out. Because of these programs' open source nature, users are free to modify the source code, provided that the modifications are made available to others.

2 WEKA

WEKA, an open source collection of data mining algorithms written in java, is a solid exploratory tool for those interested in mining their collected data (Witten and Frank 2005). Users can either use the Weka-provided interface or take advantage of

R.P. Schumaker et al., *Sports Data Mining*, Integrated Series in Information Systems 26, DOI 10.1007/978-1-4419-6730-5_9, © Springer Science+Business Media, LLC 2010

incorporating the java class libraries into their own code. While it is open source and freely distributable, Weka is covered under the GNU General Purpose License, where any changes to the software must be made freely available. Weka was developed at the University of Waikato in New Zealand and is primarily aimed at the academic community as a data mining tool. An example screenshot of the Weka tool for selected greyhound racing data is shown in Fig. 9.1.

Weka contains multiple classifier algorithms including several categories of naïve Bayesian classes, numerous fitting algorithms such as least squares, regression, neural networks and support vector machines, a handsome variety of boosting and bagging algorithms, a nice assortment of decision trees and a collection of rule-based algorithms. Aside from the classifiers, Weka also supports clustering and association rule mining.

Aside from the wealth of algorithms at your disposal, Weka also features a plethora of options such as how to partition the data between training and testing sets, options on how to filter the results and options on how to visualize the testing data. The steps for using Weka are relatively straight forward.

- Start the Weka program
- Open the file of the dataset to be mined (assuming it is in a form that Weka understands)
- Select the attributes to learn from

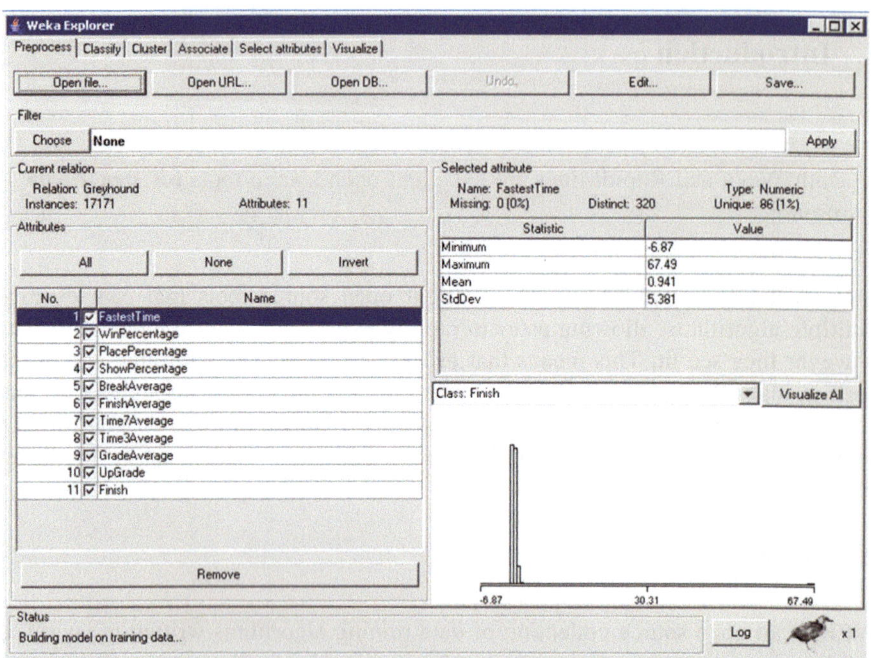

Fig. 9.1 The Weka tool for Greyhound racing data mining

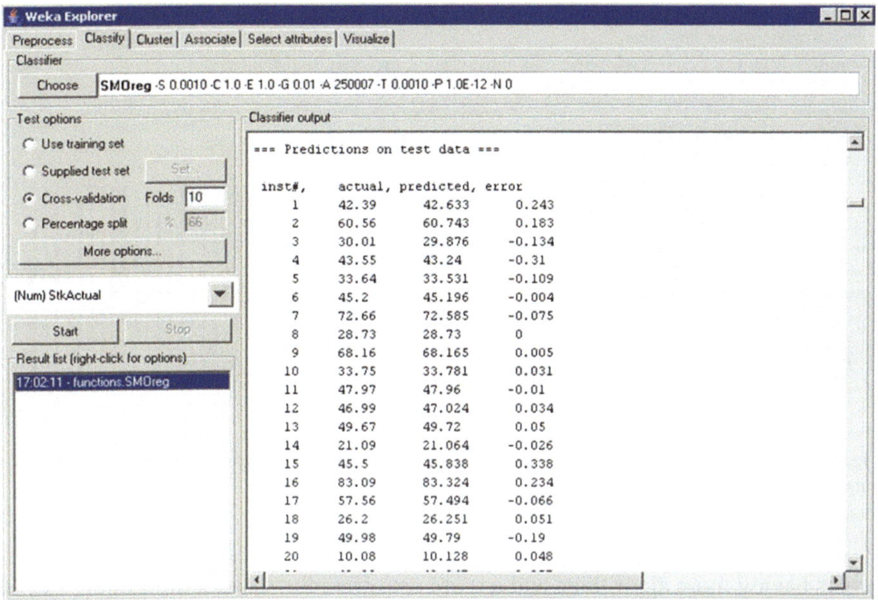

Fig. 9.2 Weka's predictive results using selected stock market data

- Select a classifier
- Select how to partition the data between training and testing
- Select what attribute to make predictions about
- Start the system

Weka will then display the predictive results to the user, as shown in Fig. 9.2.

3 Rapidminer

RapidMiner is another data mining tool, but this one is a bit unique. RapidMiner is partially open source and partially closed source. The reason for this division is because RapidMiner's core system utilizes the Weka algorithms. As a result of using Weka, Weka's GNU license requires the source and modifications to be open source. The unique aspect of RapidMiner is its focus on the frontend, in displaying the results to users. Since this part is not a part of Weka, it can be maintained as closed source.

RapidMiner comes in two varieties, one is an enterprise version in which the system will explore multiple alternatives and return the most favorable one. This commercial version is not free and is aimed primarily at larger corporations that have an extensive amount of data to mine. The other version, a community version, is available for free and performs much the same as Weka, however, RapidMiner

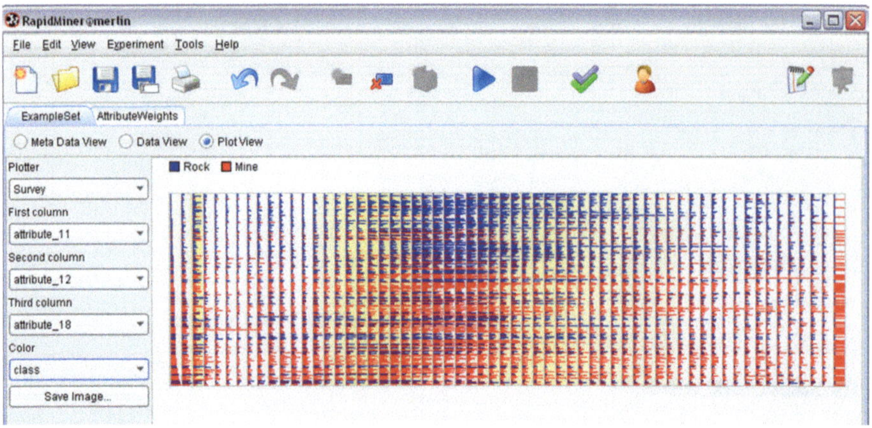

Fig. 9.3 RapidMiner visualization screenshot, courtesy of http://rapid-i.com/content/view/9/25/lang,en/

boasts having more algorithms and a more user friendly visualization interface. An example screenshot of the RapidMiner system is shown in Fig. 9.3.

4 Conclusions

Both Weka and RapidMiner are exceptional open source tools that nearly anyone with some basic computing training can use. They capitalize on an abundance of machine learning algorithms, data manipulation options, and visualization techniques. While these tools still require human direction and experimentation, it is not far off to imagine these tools as 1 day becoming one-click systems that analyze the data automatically under multiple algorithms and multiple visualization techniques and returns only those that score high on an "interestingness" scale. Both tools would be useful for effective sports data mining.

5 Questions for Discussion

1. What other open source data mining tools do you have experience with and what are their strengths?
2. How would you adopt either Weka or RapidMiner for sports data mining?
3. If you were tasked to identify "interesting patterns" in different sports data, what criteria would you use?

Chapter 10
Greyhound Racing Using Neural Networks: A Case Study

Chapter Overview

Uncertainty is inevitable in problem solving and decision making. One way to reduce it is by seeking the advice of an expert. When we use computers to reduce uncertainty, the computer itself can become an "expert" in a specific field through a variety of methods. One such method is machine learning, which involves using a computer algorithm to capture hidden knowledge from data. Machine learning usually encompasses different types of solutions, such as decision trees, production rules, and neural networks.

We compared the prediction performances of three human track experts with those of two machine learning techniques; a decision-tree building algorithm ID3 and a neural network learning algorithm. This case study helps demonstrate the systematic process required of sports and gaming data mining.

1 Introduction

Most of the applications on which machine learning has been tested are in engineering, business, or biomedical domains. These domains are complex and interesting, but the data sets used for testing were relatively clean and structured. For our research, we investigated a different problem-solving scenario called game-playing, which is unstructured, complex, and seldom-studied.

We considered several real-life game-playing scenarios and decided on greyhound racing, a complex domain that involves about 50 performance variables for eight competing dogs in a race.

R.P. Schumaker et al., *Sports Data Mining*, Integrated Series in Information Systems 26, DOI 10.1007/978-1-4419-6730-5_10, © Springer Science+Business Media, LLC 2010

Foe every race, each dog's past history is complete and freely available to bettors. This is a large amount of historical information – some accurate and relevant, some noisy and irrelevant – that must be filtered, selected, and analyzed to assist in making a prediction. This large search space poses a challenge for both human experts and machine-learning algorithms. The questions then become: Can machine-learning techniques reduce the uncertainty in a complex game-playing scenario? Can these methods outperform human experts in prediction? Our research sought to answer these questions.

2 Setting Up the Experiments

As is typical of complex problem-solving scenarios, the first, most important, and most time-consuming task is to reduce the problem's complexity. In greyhound racing, we rely on human experts' knowledge and heuristics to select a smaller set of relevant performance attributes.

The Tucson Greyhound Park in Tucson, Arizona, holds about 112 races in an average week. The park makes detailed programs available to its patrons (see Fig. 10.1). Each program contains about 15 races, with race grades varying from A (the most competitive) to D. A few special races with grades such as "M" (maiden race) are also included, but were ignored in our experiments. Each race program includes information about eight dogs, including each dog's fastest time, the dog's total races, and its number of first, second, third, and fourth place finishes. Directly below the dog's summary data, the program lists performance for the last seven of its races, which includes the dog's starting position, its position during the first turn (called the break position), its position in the second and third turns, and its finishing position. In addition, its race time and the grade of the race are recorded. The park also publishes the previous day's results. These contain information about how each dog fared, along with the payoff odds on the winning dogs.

The park also enlists the prediction services of three dog racing experts, who are not affiliated with the park. To our knowledge, their predictions are based solely on their own thinking, and the park does not compensate them for publishing their advice in the daily programs. These experts base their predictions on the same information available to any bettor in the daily program.

In a typical race, the program contains about 50 variables that might affect the outcome of that race. Among these are the condition of the track, the weather, the dog owner and trainer, and the physical attributes of each dog. Using a "multiple representation strategy" that bridges the gap between structural representation and performance representation, most of these variables in the program can be considered structural variables, from which a small set of performance-related variables must be identified (either algorithmically or by some domain experts).

Event	Trk	Dst	TC	Time	Wt	PP	Off	1/8	Str	Fin	ART	Odds	Grade	Comment
Four B Flyer								TU 7 0 0 2 0						Kennel: K&K Enterprises
								C D DQ 5 0 0 0 0						Owner: Forby Moire
Brindle F, March 3, 1992														Trainer: C. Young
08/28	TU	5/16	F	31.80	53½	1	3	1	1	5	31.98	84.96	D	Lead til Late
08/23	TU	5/16	F	32.03	53	2	1	1	1	3	32.12	8.20	D	Outfinished, Rail
08/17	TU	5/16	F	31.37	53	1	2	2	2	3	31.49	6.40	D	Game Try, Rail
08/10	TU	5/16	F	31.60	53	1	2	2	3	8	32.24	27.80	C	Steady ade
08/03	TU	5/16	F	30.94	53½	8	3	2	2	6	31.67	38.80	C	Weakened, Inside
07/28	TU	5/16	F	30.96	52½	7	5	4	5	8	32.13	17.30	C	Steady Fade, Mdtk
07/25	TU	5/16	F	31.48	53½	5	7	2	2	6	32.09	10.20	C	Weakened, Mdtk
Oh, Amanda								TU 7 0 0 2 0						Kennel: M.P. Kennels
								C D DQ 5 0 0 0 0						Owner: Jay Gaunter
Dark Brindle F, March 3, 1992														Trainer: D.Z. Phillips
08/28	TU	5/16	F	31.80	57	2	1	4	4	4	31.95	6.20	D	Closed, Mdtrk
08/22	TU	5/16	M	31.82	58½	6	4	7	8	8	32.96	13.00	D	Close Qrtrs 1st
08/16	TU	3/8	F	39.74	58	8	2	3	2	3	40.07	10.00	D	Followed the Pace
08/09	TU	3/8	S	39.71	58½	3	1	2	4	7	40.87	20.80	C	Making Move, Blkd
08/03	TU	5/16	F	30.94	57½	3	5	4	4	4	31.54	28.90	C	Midtrack Thruout
07/28	TU	5/16	F	31.39	56½	5	3	6	5	5	31.83	15.50	C	Never Prominent
07/25	TU	5/16	F	31.42	57½	7	3	8	6	6	32.52	10.80	C	Blkd Much 1st Trn
Thursday's Doll								TU 7 0 0 2 0						Kennel: Charlie's Pack
								C D DQ 5 0 0 0 0						Owner: Al Hughes
Red Brindle F, March 3, 1992														Trainer: Raymond N. Peter
08/28	TU	5/16	F	31.67	68½	3	2	3	3	2	31.74	3.50	D	Led Til Late
08/23	TU	5/16	F	32.03	69	5	5	5	6	5	32.44	5.10	D	No Threat, Mdtrk
08/14	TU	5/16	F	31.25	69½	4	2	4	4	4	31.73	8.10	D	Threat 1st, Blkd
08/08	TU	5/16	M	31.81	69½	7	6	2	2	3	32.21	4.70	D	Chased the Winner
08/03	TU	5/16	F	31.23	69½	8	2	4	6	6	OOP	3.50	D	Stumbled, Backstr
07/27	TU	5/16	F	31.18	69½	7	5	2	2	2	31.41	8.50	D	Chased the Winner
07/20	TU	5/16	F	31.13	70½	1	1	8	8	8	32.45	5.00	D	Stumbled, 1st Trn

Fig. 10.1 Sample greyhound racing program

As is typical of complex problem-solving scenarios, the first, most important, and most time-consuming task was to reduce the problem's complexity by pruning the problem space: deciding on a smaller set of relevant performance attributes. The inability of the machine learning algorithms to analyze the initial noisy and fussy problem domain prompted us to rely on human experts' knowledge and heuristics.

As in horse racing prediction, the selection of performance attributes in our research was mainly based on the opinions of frequent bettors, track experts, and the management of the park. These experts succinctly identified performance variables that they thought were the most crucial in predicting winners. Their domain knowledge helped to reduce the problem space significantly. In total, the experts suggested ten performance variables:

- Fastest time: the fastest time in seconds for a 5/16 mile race
- Win percentage: the number of first places divided by the total number of races
- Place percentage: the number of second places divided by the total number of races
- Show percentage: the number of third places divided by the total number of races
- Break average: the dog's position during the first turn (averaged over the seven most recent races)
- Finish average: the average finishing position over the previous seven races
- Time 7 average: the average finishing time of the seven most recent races
- Time 3 average: the average finishing time of the three most recent races
- Grade average: the average grade of the seven most recent races the dog competed in
- Up grade: weight given to a dog when dropping down to less competitive race grade

Since the dogs compete against each other directly in a win-lose scenario, we manipulated the data so that the values were relatively scaled. This means that after making all calculations (such as averaging a dog's finish times), we than assigned the lowest value in one single race to be zero. The other values were scaled to reflect their difference from that lowest value.

We used two-thirds of the data training and one-third or testing. The training stage consisted of 200 races, for a total of 1,600 classified greyhounds. Testing consisted of an additional 100 races, or 800 new cases.

3 Testing ID3

ID3 adopts a divide-and-conquer strategy for classification. Its goal is to classify mixed objects into their associated classes, based on the objects' attribute values. In our application, each greyhound can be classified as either a winner or a loser, and described by a set of attributes such as the number of races it has won, its fastest time, and so on.

ID3 is designed to deal with both categorical attributes and continuous values. For categorical attributes, attribute values can be easily enumerated. For continuous values, ID3 performs a sweeping analysis of entropy reduction for all possible partition points (between any two consecutive values) and selects the partition that reduces entropy the most. In essence, the algorithm performs a binary partition. For our system, we developed an ID3 program that adopted this original design.

We also developed a version of the ID3 algorithm that performed ternary partition for continuous variables. This variant of the ID3 algorithm first sorted the continuous values in ascending order. It then marked the "clean" classes on the two ends of the sorted list as unique classes. Other values in the middle of the list were mixed and remained to be classified using other attributes.

In our implementation, it took only several seconds to build a decision tree from 1,600 training cases. ID3's result was simple to understand, and it was easy to trace the classification decisions.

The attribute break average occupied the root node. Branches represented the up grade, the show percentage, the fastest time, the grade average, the win percentage, time 7 average, finish average, time 3 average, and place average. The decision tree created by ID3 using ternary partition is shown in Fig. 10.2. The first five

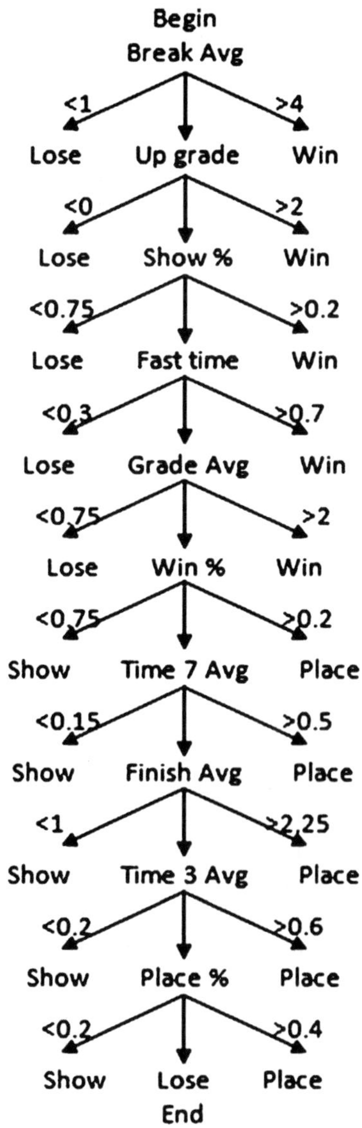

Fig. 10.2 An ID3 decision tree

attributes were useful for deciding first-place winners. Other attributes were helpful for deciding place (second place) and show (third place), which were not considered in our analysis.

4 Testing the Backpropagation Neural Network

The number of units in the backpropagation input layer is determined by the number of independent variables in the data set. Similarly, output units represent the application's dependent variables (for prediction problems) or classes (for classification problems). For greyhound racing, the input layer consisted of the various dog racing attributes that could affect the outcome of winning or losing in the output layer. Given a greyhound's past history in racing and other important features, the network should be able to predict the dog's performance in a given race.

If all possible inputs and outputs are shown to a backpropagation network, along with proper topology and training epochs, the network will likely find a set of weights that maps the inputs onto the outputs correctly.

5 The Results

In gaming prediction different data mining techniques may achieve different levels of performance in accuracy and payoffs. Neural networks are able to obtain high payoffs for long shots.

Evaluating the performance of different techniques was simply a matter of computing the monetary payoffs. For each dog that an algorithm or an expert predicted to be winner, we bet \$2.00. If the algorithms predicted more than one winner, we bet on all predicted winners; if no winner was picked, we placed no bet. Using the payoff odds given in the result sheets, we were able to compute the final payoffs after betting on 100 races. For performance comparison, we only considered the payoff for first-place winners; we did not consider more elaborate betting systems such as place (your greyhound must finish either first or second); show (your greyhound must finish either first, second, or third), quiniela (two greyhounds you bet on must finish first and second), and so on. It could be argued that the payoff odds are not necessarily correct, since each bet on a particular dog will change the payoff odds for that dog. We contend that since there are numerous bets on any given dog, the odds would not change significantly with the additional bet on a particular dog.

Table 10.1 summarizes the predictions by the experts, ID3, and the backpropagation network, and their final payoffs or 100 races.

Table 10.1 Predictions and payoffs for 100 races

Technique	Correct	Incorrect	Did not bet	Payoffs ($)
Expert 1	19	81	0	-$71.40
Expert 2	17	83	0	-$61.20
Expert 3	18	82	0	-$70.20
ID3	34	50	26	$69.20
Backprop.	20	80	0	$124.80

Among the three track experts, the best predicted the winners in 19 races, but predicted 81 races incorrectly. Or all experts, the final payoffs after betting on 100 races ($2 per race) were negative: -$71.40, -$61.20, and -$70.20. That is, when following the experts' predictions, for a total of $200 bet on 100 races, a bettor lost about $70.

ID3 correctly predicted winners for three races, but incorrectly predicted 50 outcomes. In 26 races, ID3 did not predict any winners, and thus no bet was placed. The final payoff for ID3 predictions was $69.20 for the 100 races. Compared with the most successful expert, ID3 had a more accurate and sometimes more conservative prediction record. The monetary gain for ID3 was mainly obtained from the high payoffs for a few races – $24.40 for race 31, $41.20 for race 59, and $26.80 for race 83. The experts' best payoff was $11.40. By analyzing the greyhound attributes objectively, ID3 identified long shots and thus realized significant monetary gain.

Our backpropagation network consisted of a simple three-layer design. We tested hidden units of 15, 20, 25, 30, and 35, respectively. The network of 25 hidden units consistently made better predictions than other topologies. The resulting optimal backpropagation network predicted 20 races correctly, and mis-predicted 80 races. Even though the number of correctly predicted races was about the same as for the human experts, their monetary payoffs were very different. The backpropagation network gained $124.80 for the 100 races. As with the ID3 results, the backpropagation algorithm obtained high payoffs for several long shots – $78.00 for race 10, $30.20 for race 37, $30.00 for race 64, and $26.80 for race 83.

Two sample t-tests revealed that the backpropagation network outperformed the best expert (and the other experts) in monetary payoffs at the 10% significance level ($P = 0.084$). However, ID3 did no outperform the experts at a statistically significant level ($P = 0.15$) and there was no statistically significant difference in payoffs between backpropagation and ID3 ($P = 0.68$).

6 Conclusions

In this experiment, especially during the early stage, experts' heuristics for refining the initial performance variables were critical to the success of the machine learning algorithms. The 50-plus candidate variables were reduced to 10 of the most relevant parameters, which effectively reduced the problem space for this complex and noisy domain. Some variables were also the results of computations, such as time 3 average, time 7 average, and so on.

In terms of prediction accuracy and monetary payoff, both the backpropagation network and ID3 performed better than human experts. Both algorithms appeared to be more robust than humans in their ability to analyze the large data set of racing data objectively and reach unbiased conclusions. This characteristic is particularly evident from both algorithms' convincing prediction of numerous long shots that human experts failed to identify. In comparing the two algorithms, ID3's decision tree output was more understandable than that of backpropagation. In general, ID3 also predicted more conservatively than other approaches. The backpropagation network, on the other hand, was more computationally expensive (and thus very slow), but made excellent predictions of long shots.

7 Questions for Discussion

1. What did the machine-learning algorithms "see" that the human experts missed?
2. The neural network had better accuracy and the ID3 algorithm a better payoff. Can you think of any ways to combine the two?
3. Do you think the same approach can be applied to thoroughbred or harness racing? How about track and field events or ball games?

Chapter 11
Greyhound Racing Using Support Vector Machines: A Case Study

Chapter Overview

In this chapter we investigate the role of machine learning within the domain of Greyhound Racing. We test a Support Vector Regression (SVR) algorithm on 1,953 races across 31 different dog tracks and explore the role of a simple betting engine on a wide range of wager types.

1 Introduction

Greyhound racing is recognized as one of the nation's largest spectator sports. According to the American Greyhound Track Operators Association, it is currently legal in 16 states: Alabama, Arizona, Arkansas, Colorado, Connecticut, Florida, Iowa, Kansas, Massachusetts, New Hampshire, Oregon, Rhode Island, South Dakota, Texas, West Virginia and Wisconsin. Other states do not hold live racing but offer simulcasts (broadcasts of remote races) for betting.

Greyhound racing has a following that parallels horse racing. Many of the same elements in animal athleticism exist in both sports and also the betting on these races. The key to betting is determining how to systemically predict the winners and combinations of winning bets. Bettors must carefully read all the information on the race card and gather as much information about the dogs as possible. Bettors will examine the dogs and their physical conditions, how they have shown in past races, their breeding and bloodlines, their assigned grades for how well they perform, as well as their odds against the dogs they will race against. Weather also plays a role as some bettors rely on the inside traps (part of the track) during wet races. Even

R.P. Schumaker et al., *Sports Data Mining*, Integrated Series in Information Systems 26, DOI 10.1007/978-1-4419-6730-5_11, © Springer Science+Business Media, LLC 2010

weights of the dogs come into play. Lighter dogs and the longshots tend to get bumped and pushed out of the running in the first couple of urns.

Our research motivation is to build upon prior machine learning techniques in the domain of Greyhound Racing and more closely examine the effect of longshots on racing prediction and payouts. We further examine the impact of non-traditional wager types and more robustly examine their resulting accuracy, payouts and return on wagers.

2 Relevant Literature

Research on predicting race outcomes can generally be broken into three distinct areas: mathematical, psychological, and data mining techniques.

Research on predicting race outcomes can generally be broken into three distinct areas: mathematical, psychological, and data mining techniques. Mathematical areas include Harville formulas, which are a collection of equations that establish a rank order of finish by using combinations of joint probabilities. Other mathematical formulas include the Dr. Z System, where a potential gambler waits until 2 min before race, select those dogs with low odds and bet Place (i.e., the dog will finish in second place or better) on those with a win frequency to place frequency greater than or equal to 1.15 and bet Show (i.e., the dog will finish in third place or better) on those with a win to show frequency greater than or equal to 1.15. This successful system received considerable attention from both academics and gamblers alike.

In psychological methods, perhaps the best known method for selecting a dog is the longshot bias. Arrow-Pratt theory suggests that bettors will take on more risk in order to offset their losses. Gamblers tend to favor the low probability, high payout combinations for luck, entertainment or desperation, but it has never been found to yield sustainable positive returns on combination bets. Another point of view is to place betting in terms of Stock Market Efficiency. In this approach it is argued that betting on favorites should be as profitable as betting on longshots.

In data mining methods, simulations can be used to predict outcomes. This method has been used with some degree of success in yacht racing and the thoroughbred industry. However, simulated data does not address the complexities involved with large numbers of varying parameters. Another technique in data mining is to use statistical learning methods. These systems are better able to generalize the data into recognizable patterns. One of the more recent methods in statistical learning is Support Vector Regression (SVR), which is a variant of Support Vector Machines (SVM). Both SVM and SVR are at their essence classification algorithms that seek to maximally classify high dimension data while minimizing their fitting error. SVR differs in the respect that the hyperplane used to divide the sets can be used as a regression estimator and can return discrete values instead of categories. This technique was used in a similar context to predict stock prices from financial news articles (Schumaker and Chen 2008).

In a prior study of greyhound races, Chen et al. tested an ID3 and Back Propagation Neural Network (BPNN) on ten race performance-related variables determined by domain experts on 100 races at Tucson Greyhound Park (Chen et al. 1994). From their work they made binary decisions as to whether the greyhound would finish first based on its historical race data. If a dog was predicted to finish first, they would make a $2 wager. The ID3 algorithm resulted in 34% accuracy and a $69.20 payout while the BPNN had 20% accuracy and a $124.80 payout. This seeming disparity in decreased accuracy and increased payout is justified with the argument that the BPNN was more successful in selecting longshot winners, hence accuracy would suffer but the long odds would result in the higher payouts. By comparing their machine learning techniques to track experts, the experts managed a much more disappointing 18% accuracy and a payout loss of $67.60.

In a follow-up study that expanded the number of variables studied to 18, Johansson and Sonstrod used a similar BPNN but also investigated the effect of more exotic wagers such as Quiniela (i.e., selecting the first two dogs to finish in any order) and Exacta (i.e., selecting the first two dogs to finish in order) (Johansson and Sonstrod 2003). Their study on 100 races at Gulf Greyhound Park in Texas found 24.9% accuracy for Wins and a $6.60 payout loss. This seemingly better accuracy and worse payout than Chen et al. would imply that it was either the additional variables or too few training cases (449 as compared to Chen's 1,600) that harmed their ability to capture longshots. However, their exotic wagers did better. Quiniela had 8.8% accuracy and a $20.30 payout, while Exacta had 6.1% accuracy and $114.10 payout.

From our review we propose the following research questions. How accurate is a machine learning method in predicting Greyhound race outcomes? How profitable is the same system? How will the addition of exotic wagers affect system accuracy and profitability?

3 Research Methodology

Racing data can be extracted automatically from web site such as trackinfo.com and analyzed with data mining tool such as Weka.

To address these research questions, we built the AZGreyhound system as shown in Fig. 11.1.

The AZGreyhound system consists of several major components: the data gathering module, the machine learning part, the betting engine and the evaluation

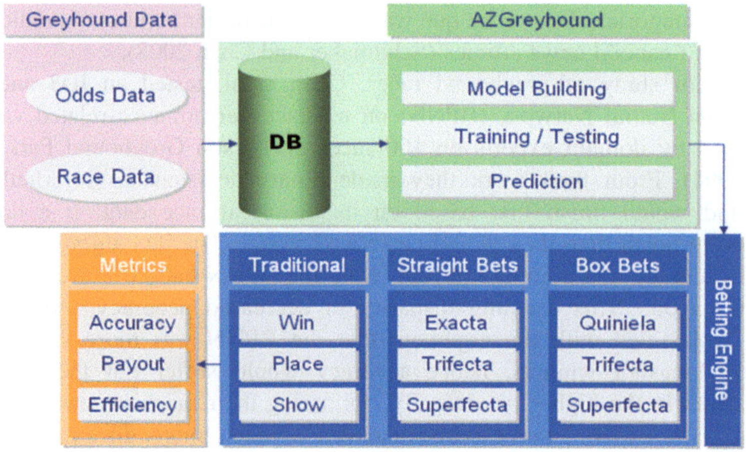

Fig. 11.1 The AZGreyhound system

metrics. The odds data is the individual race odds for each wager type (e.g., Win, Place, Show, etc.).

The race data are features gathered from the race program. Each race program contains a wealth of data. There are generally 12 races per program where each race has six to eight dogs. The usual number of dogs per race is eight, but some dogs may scratch (i.e., not race) which can lower the field of competition. Also within the race program, each dog has the results from the previous five races. There is some dog-specific data within the race program such as the dog's name, color, gender, birthday, sire, dam, trainer and kennel. Race-specific race information includes the race date, track, fastest time, break position, eighth-mile position, far turn position, finish position, lengths won or lost by, average run time, grade of race, track condition and racing weight.

Once the system has been trained on the data provided, the results are tested along three dimensions of evaluation: accuracy, payout and efficiency. Accuracy is simply the number of winning bets divided by the number of bets made. Payout is the monetary gain or loss derived from the wager. Efficiency is the payout divided by the number of bets which is used for comparative purposes to the prior studies.

The Betting Engine examines three different types of wagers: traditional, straight bets, and box bets. In traditional wagers, bettors speculate on whether a dog will win, place or show. If betting on a Win, the bettor receives a payout only if the selected dog comes in first place. If betting on Place, the bettor receives differing payouts if the selected dog comes in either first or second place. If betting on Show, the bettor receives differing payouts if the selected dog comes in first, second or third place. In straight bets, bettors consider the finish placement of multiple dogs through Exacta, Trifecta and Superfecta wagers. In Exacta, the bettor is trying to predict which two dogs will come in first and second place respectively. In Trifecta, the job is made more difficult by trying to guess the placement of the first three dogs in order. Superfecta is even more difficult where the bettor is trying to determine which four dogs will cross the finish line in order. Box bets simplify the selection process by taking finish order out of the equation. In essence you are betting on every combination of finish between the selected dogs. This makes box betting a more expensive wager.

3.1 Data Acquisition

To perform our experiment, we automatically gathered data from www.trackinfo.com, which consists of daily race programs and odds charts for all US Greyhound tracks in operation at the time of this study. Some tracks contained multiple daily programs incorporating both afternoon and evening racing. We eliminated schooling races from the data, as they simply assign race grades to greyhounds and do not contain the full amount of race data. Once the data was gathered, it is parsed to obtain specific race data and sent to AZGreyhound for prediction.

For our collection of greyhound races we chose a study period of January 7 through March 7, 2007. Prior studies used only one racetrack, input their data manually and had small datasets. Chen et al. (1994) used 1,600 training cases from Tucson Greyhound Park, whereas Johansson and Sonstrod (2003) used 449 training cases from Gulf Greyhound Park in Texas. Our study differs by automatically gathering race data from multiple tracks. In all, we gathered 41,473 training cases covering 7,760 races. However, we were only able to use 1,953 races because race programs list the race results of the prior five races and we needed data on the prior seven races. This data incorporated 7,163 dogs from 31 different tracks, however, seven of the tracks provided the bulk of data as shown in Table 11.1.

Table 11.1 Number of races gathered from various tracks

Track	# Races
Caliente	1,408
Raynham/Taunton	1,323
Lincoln	1,174
Wichita	962
Melbourne	898
Tucson	788
Hinsdale	700
All others	507

3.2 Support Vector Machines Algorithm

Using Chen et al. (1994) as a guide, we limited ourselves to ten race variables over the most recent seven races for each greyhound:

- Fastest Time – a scaled difference between the time of the race winner and the dog in question, where slower dogs experience larger positive values.
- Win Percentage – the number of wins divided by the number of races
- Place Percentage – the number of places divided by the number of races
- Show Percentage – the number of shows divided by the number of races
- Break Average – the dog's average position out of the starting box
- Finish Average – the dog's average finishing position
- Time 7 Average – the average finishing time over the last seven races
- Time 3 Average – the average finishing time over the last three races
- Grade Average – the average racing grade of the dog (A–D) of the last seven races
- UpGrade – additional points given to a dog racing in a less competitive grade (e.g., +3 points if the most recent race was a better grade, +2 points if the drop in grade was two races back, +1 if the drop was three races ago, 0 otherwise)

For the machine learning module we implemented Support Vector Regression (SVR) using the Sequential Minimal Optimization (SMO) function through Weka (Witten and Frank 2005). SVR allows for discrete numeric prediction instead of classification. We also selected a linear kernel and used ten-fold cross-validation.

4 Results

Within traditional wagers the Show bet appeared the best in accuracy and payoff. This stems from AZGreyhound picking greyhounds with longer odds and subsequently the higher payouts. For exotic wagering, Superfecta Box had highest payout.

To answer our first research question, on how accurate is a machine learning method in predicting Greyhound race outcomes, we performed a sensitivity analysis by varying the SVR parameter cutoff's from 1.0 to 8.0 on Win, Place and Show as shown in Fig. 11.2. Show was most accurate and Win was least accurate as expected because betting Show will provide a positive return if the greyhound Wins, Places or Shows. Accuracy decreases when cutoff increases and peaks at about 1.7 for Win.

To answer our second research question of how profitable is the same system, we performed a sensitivity analysis of system payout as shown in Fig. 11.3. Show had the best payouts and peaked at Cutoff 4.2 with a $7,341.95 payout on 8,289 bets. Win, Place and Show all had positive payouts between cutoffs 1.6 and 4.6.

To answer our third research question, how will the addition of exotic wagers affect system accuracy and profitability, we analyzed the addition of straight and box wagers to payout as shown in Fig. 11.4.

From this figure, we can see that Superfecta Box had the highest payout at $41,517.37 at cutoff 4.1 before leveling off. By contrast, Exacta appears to be a poor bet, with AZGreyhound losing money for cutoffs above 1.6. Both Superfecta Box and Trifecta Box garnered substantially higher payouts than the other exotic

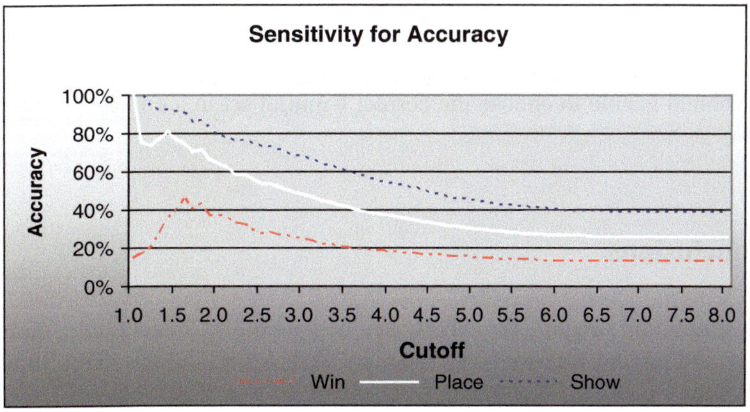

Fig. 11.2 System accuracy for traditional wagers

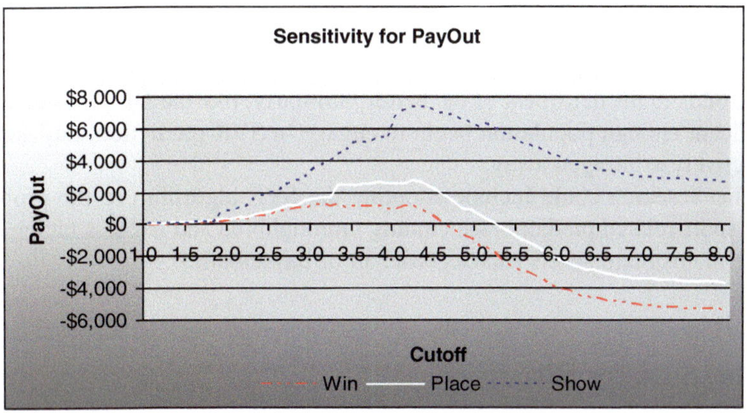

Fig. 11.3 System payout for traditional wagers

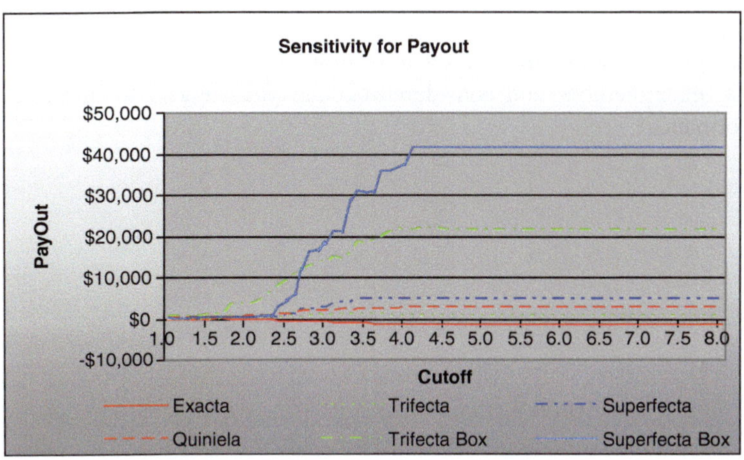

Fig. 11.4 System payout for exotic wagers

wagers for two reasons. First, since the odds of correctly selecting the Trifecta and Superfecta combinations are markedly low, these types of wagers will inherently have higher payouts. Second, in spite of these low odds of correct selection, AZGreyhound is able to choose the correct winning set in a consistent manner.

5 Conclusions

Within traditional wagers the Show bet appeared the best. Show had higher accuracy and payout followed by Place and Win for all cutoffs. This stems from AZGreyhound picking greyhounds with longer odds and subsequently the higher payouts. For exotic wagering, Superfecta Box had the highest payout for cutoffs above 4.1. This is also the result of AZGreyhound able to capitalize on the longer odds more accurately than random chance alone.

While this system demonstrates a marked promise of better prediction, the reader should be cautioned that the act of making large bets on races will change the race odds to the detriment of the bettor. Similarly, like the Dr. Z system, should a significant enough population begin to engage in SVR prediction, any gains will be effectively arbitraged away.

Further research could include adopting the SVR algorithm to the problem of similar sport-related predictions including thoroughbred and harness racing as well as more mainstream sports such as baseball or basketball.

6 Questions for Discussion

1. How could a system like this be ported to similar sports such as thoroughbred and harness racing?
2. Balancing between accuracy and payout can be tricky. How would you suggest balancing between them with a betting strategy?
3. What other methods and considerations can you suggest to improve betting performance?

Chapter 12
Betting and Gaming

Chapter Overview

How is it that sports and gambling co-exist so easily together, yet can cause so many problems? We explore the relationship between sports and gambling from a historical perspective and describe the ways that some organizations are trying to keep a safe distance between the two.

1 Introduction

Over the years, numerous problems have crept up between sports and gambling. Most of the time this activity has had a criminal bend, where matches are fixed, players underperform, and corrupt officiating occurs. These occurrences have led to reforms, not only in the sports environment itself, but also in the legal system. Unfortunately, gambling and sports have a long history together, and the process of isolating the two is largely reactionary in nature.

2 The Effects on Gambling on Sports

> Gambling and sports have always had an uneasy relationship. Sometimes the effects of gambling have spilled over into sports in a detrimental way; from the 1919 Chicago White Sox fix and Pete Rose's baseball betting, to the corrupt NBA referee Tim Donaghy and several international soccer scandals.

Gambling and sports have always had an uneasy relationship. Sometimes the effects of gambling have spilled over into sports in a detrimental way.

R.P. Schumaker et al., *Sports Data Mining*, Integrated Series in Information Systems 26,
DOI 10.1007/978-1-4419-6730-5_12,

The 1919 Chicago White Sox were baseball's tipping point of this spill-over effect. For years, crooked behavior and gambling interests had been creeping into baseball, damaging its public integrity. However, it wasn't until 1919 that a fix so large occurred that the problem of gambling in sports needed to be addressed. Dubbed the Black Sox scandal by the press, bettors were allegedly able to influence eight Chicago players into throwing the 1919 World Series. Following the public outcry, baseball owners hired a tough commissioner, Judge Kenesaw Mountain Landis, who tried to restore integrity and restore public trust by banning the eight players for life as well as enacting tough rules regarding baseball and the prohibition of gambling. While these steps are credited with restoring baseball, they were not able to prevent further instances. Nearly 70 years following the Black Sox scandal, another scandal related to gambling was brewing and this one involved potential hall of famer player/manager Pete Rose. In this case, Rose was allegedly wagering on his team to win games and tried to use this argument in his defense. However, the Baseball Commissioner's Office saw the situation as a violation of baseball's rules regarding gambling and placed Rose on baseball's ineligible list, effectively banning him from the game.

Basketball has also had problems with gambling and some of the aftershocks of which are still reverberating through the NBA. In this case, NBA referee Tim Donaghy had incurred substantial gambling debts and was contacted by mob officials to control the spread of games. During this time Donaghy was in charge of officiating hundreds of games, and was so adept at his deception that the NBA did not find out until they were contacted by the FBI in connection with the mob's involvement. In hindsight, the clues were there, however the ability to decipher those clues was not. By looking back through the records to find these clues, the NBA found that in 57% of the games Donaghy was officiating, the teams beat the sportsbook spread, which is calculated to be a 0.001% occurrence by chance alone. Furthermore, it was found that in games that Donaghy was officiating, exceptionally large bets were being made, which could have been a signal that something was wrong. Donaghy later pleaded guilty to his part in the gambling scandal and was sentenced to federal prison. For its part the NBA has been attempting to clean up its image in the aftermath, however, new allegations have surfaced of more corrupt officials and charges that the league itself was involved in tampering with games.

American sports are not the only ones prone to gambling influences. In many countries and federations, soccer is becoming a haven of illicit gambling activity and match fixing. In Italy, a 2006 scandal rocked the Italian premier league and more specifically the team of Juventus. It was found that team managers were actively engaged in selecting referees that would bias Juventus in matches (Delaney 2006). This scandal had far reaching effects and eventually led to multiple resignations from Italian soccer's governing body.

In 2005 a German official, Robert Hoyzer, came forward and admitted to match fixing in at least six German-league games (Biehl 2005). Approached by Croatian mobsters with the lure of easy money, Hoyzer agreed to tip the matches into their favor. While Hoyzer received condemnation for his role in the affair, he received a paltry amount, close to €60,000, compared to his Croatian associates who made €2 million in profit.

Recently in 2009, European soccer as a whole has been feeling the effects of yet another gambling scandal. This scandal involved at least 17 people, mostly players, officials and trainers, across nine countries. While the full details are still unraveling, it is suspected that match-fixing occurred in several Champions League games as well as many more in the rival Europa League (Hughes and Pfanner 2009).

China has also been experiencing their own series of gambling problems in soccer. The Chinese Football Association (CFA), the body responsible for oversight of Chinese soccer, has come under fire recently for a myriad of problems including bribery of officials and match-fixing (China Daily 2009). However, the problems in Chinese soccer appear systemic and not likely to change even given that the head of the CFA was taken away by police. It is reported that the CFA works closely with law enforcement to attempt to prevent such occurrences, however, corruption in the league is widespread. Similarly, Colombia is having to deal with these issues as well. In 2009, a scandal erupted casting a pall on game integrity where a highly suspicious call was made by officials which ended up tipping the balance of a game (Casino City Times 2009). League officials are reviewing the circumstances and as of yet have not released their findings.

3 Sportsbooks and Offshore Betting

Sportsbooks in the United States are highly regulated entities. Following the years of corruption and problems during the early part of the twentieth century, laws were enacted to oversee sportsbooks and attempt to keep the criminal element out from the mix. As a result of these federal laws, sportsbooks are only in existence in the state of Nevada. Although the states of Delaware, Montana and Oregon are also permitted to operate sportsbooks because of how the federal law was written, none of these additional states have done so. However, professional sports teams have felt that these laws did not go far enough and as a consequence for Nevada's hosting of sportsbooks and the uneasy history between sports and gambling, Nevada does not have any professional sports teams. It is felt that by keeping a geographic distance between the sportsbooks and professional teams, a sort of barrier has been built between them. Recently though, Delaware has been looking into legalizing sportsbooks as a way to generate additional revenue for the state. This move has quickly brought resistance from the four major professional sports bodies, baseball, football, basketball and hockey as well as the collegiate sports body, NCAA (McCarthy and Perez 2009). In the case against Delaware's plans, it is argued that given Delaware's proximity to existing professional sports teams, the temptation for gamblers to influence games through match-fixing and beating spreads is too great. However, critics note that professional sports are less than pure in their intentions, given that several teams have played games in casino venues and/or allowed their logos to be used by state lotteries.

Collegiate sports have been the targets of illegal activity as well. One of the more famous examples occurred in 1951 when it was discovered that over the prior 5 years, approximately 86 basketball games were affected by instances of point shaving. This scandal involved players, coaches, alumni and gamblers alike, leading to 20 convictions and bans from the NBA (Merron 2006). Boston College's basketball program was similarly marred during the 1978–1979 season when a New York mobster convinced one of the BC players to point shave nine separate games during the season. Boston College again got into point shaving trouble in 1996, this time with their football program, as details emerged that 13 players engaged in point shaving activity.

Outside of the United States, online sportsbooks have emerged as an alternative to the Nevada sportsbook system. However, these offshore sportsbooks are largely unregulated and operate under laws of different jurisdictions. While some offshore sportsbooks are legitimate extensions of existing businesses, many others border on the fraudulent by refusing to provide winnings, charge exorbitant taxes/fees or provide purposefully poor customer service as a way of discouraging patrons from collecting their winnings. As a result of these problems, the United States passed legislation where citizens are not permitted to engage in offshore gaming.

4 Arbitrage Methods

When betting on sporting events, depending upon the type of sport and gambling devices available, there are a variety of arbitrage techniques available where the bettor can "tip the odds" so to speak, in their favor. The term arbitrage can be defined as simply taking advantage of an imbalance in an equation, whereas in this case, the equation could be the tote board or some form of information present in the market. Some of these arbitrage techniques are mathematical in nature. As mentioned in an earlier chapter, the Dr. Z system is a collection of mathematical probabilities that leverage racing odds. By correctly identifying racing situations in which the favored thoroughbred/greyhound/etc's probability of winning does not match the odds provided, an astute gambler could take advantage of this market imbalance. However, it is important to note that following the release of the Dr. Z system, there were a sufficient number of gamblers using it, which ironically returned the markets to equilibrium and voided the imbalances. Because as a bet is placed in a parlay system, the odds shift to reflect the wager. As more wagers are created for the favorite, the odds on the favorite move lower which directly impacts the potential return.

Aside from the mathematical approach, machine learning techniques can be used as well. As described earlier, neural networks are an astute learner of patterns in the data. These systems can identify combinations of variables that can lead to winning wagers. By feeding sports data into a neural network, the system will self-adjust itself in an iterative process, attempting to flex and mold itself as best it can to the data presented. While neural networks do not identify arbitrage opportunities like the Dr. Z system, they instead create these opportunities through their predictions. However, because of how the parlay system is constructed, neural networks suffer from the same problem. If a bettor uses the neural network system to identify an animal to wager on, the amount of wagers placed on that animal will decrease the odds and the eventual payoff. This leads to a tradeoff between accuracy and payout. While the system can identify winners more accurately than chance alone, the act of wagering alone will influence the odds in a detrimental way. To put it another way, while you may be winning using this type of a system, you are winning far less in returns.

The other notable machine learning technique is to use support vector machines to learn from the data. This method, as described earlier, is more accurate than neural networks in most applications. This method is somewhat similar to the neural network except that data points are plotted within a high dimensional environment and a hyperplane is calculated to weave through the data points and attempt to minimize the fitting errors. This hyperplane can then be used for prediction purposes. Like neural networks, a support vector machine does not recognize arbitrage opportunities, it instead creates them and like the neural network, support vector machines also have the same weakness when applied to a parlay system. The more money that is wagered on a chosen event or target will decrease the odds and lead to lower returns.

5 Cautions and Gambling Pitfalls

While systems and methods have been constructed, detailed and described in this book, it is with caution that gamblers should use this knowledge. As already mentioned, the simple act of using these systems in a parlay environment to place a wager will decrease the returns. This in turn leads to more wagering in order to "make bank." These systems are not foolproof and do not guarantee success. The more wagering performed increases the bettors exposure to risk. If you feel as though you may have a gambling problem please contact your local gambling problem helpline.

6 Conclusions

Will there be a time when gambling and sports do not tangle with one another and dance their weave of destruction? For the near future, this idea of complete separation is doubtful. What once was used as a measure of separation, geographic distance between organized gambling and sports, has quickly come under fire. With states other than Nevada and even offshore organizations attempting to crack open the lucrative sportsbook markets, geographic distance in the physical sense will not matter soon. Even more so, the advancement of using computer networks for sportsbook operations has nullified the idea of geographic separation altogether by effectively placing a sportsbook on the screen of every networked device. While laws, rules and legal challenges are issued to curb the dangers posed, one thing is certain – more instances of compromised matches, corrupt officiating and point shaving will occur in the future. How much of it will occur and the extent is anyone's guess.

7 Questions for Discussion

1. Can you think of any way to separate the wagering markets from sports that is different from today's approaches?
2. Some have argued that the addition of more sportsbooks, outside of Nevada, will not affect current circumstances. What is your opinion on this debate?
3. What other arbitrage opportunities are available in the sports betting markets?

Chapter 13
Conclusions

Chapter Overview

Over the next several years, sports data mining practices will be faced with several challenges and obstacles. The most obvious of which is to overcome the years of resistance by the members of sporting organizations that would rather stick with a traditional way of doing things. Aside from the challenges that are faced, sports data mining currently sits at a pivotal junction in history with many opportunities just waiting to be grabbed. Some avenues of opportunity will be pursued quickly, while others may take years or decades to become fruitful. In any event, sports data mining today is still in its infancy. While some first steps were made with pioneers such as Dean Oliver and Bill James, the next few years will become a transition period as the technology begins to mature within the sporting community and become more commonplace. New metrics, algorithms and ways of thinking will begin circulating themselves as the field enters puberty and begins to mature. The coming decades will be fascinating to watch.

1 Sports Data Mining Challenges

There are many challenges within the domain of sports data mining that needs to be addressed. The first of which is that not many sporting organizations understand or make use of advanced sports data mining techniques. Oftentimes there is a resistance to change that is firmly ingrained in certain sports such as baseball that likely stems from the old axiom "if it ain't broke don't fix it." However, sporting organizations should realize that those that do embrace these sophisticated instruments, generally perform better.

Secondly, the individual sports organizations that have recognized the potential competitive advantages of data mining systems, typically contain their results in-house and do not share either the technologies or lessons learned with fans or

R.P. Schumaker et al., *Sports Data Mining*, Integrated Series in Information Systems 26, DOI 10.1007/978-1-4419-6730-5_13, © Springer Science+Business Media, LLC 2010

peer groups. While this approach could be considered selfish, there are other sports organizations that take a different approach and store all game data in a central sport-related repository where individuals and teams have equal access. Both approaches have their respective advantages, however, the missing pieces are hybrid approaches where a significant amount of material is housed collectively and teams are still free to exploit any advantages found therein. We see the beginning of such a hybrid approach for certain sport-related interest groups.

The Australian Institute of Sport (AIS) has recently unveiled two initiatives designed to effectively store data from various sports (Lyons 2005). The first of which is the creation of a digital repository to store various sport-related multimedia and data files. This centralized repository system will allow players and teams to access any relevant data. The second initiative provides data mining techniques on the data repository in order to gather new insight into obvious patterns. We would suggest that further knowledge could be extracted from such a repository and that individual teams and players should take advantage by implementing their own proprietary tools in the pursuit of a competitive advantage.

2 Sports Data Mining Audience

There are many possibilities for organizations and individuals who are interested in sports data mining. In particular, we believe the following audiences will benefit from this book and the new sports data mining approach:

- Executives, managers, and scouts from different professional sports organizations could gain from systematically collecting and analyzing various player and opponent data in order to maximize their investment.
- Athletic department directors, administrators, coaches, managers, and trainers in schools and universities could use the data mining approach to better monitor and train their student athletes and identify areas for improvement.
- Sports governing body, associations, and relevant government departments will benefit from continuous, systematic monitoring and analysis of team and player activities to identify potential irregularities and frauds.
- Players at professional and competitive amateur teams can better understand their strengths and weaknesses and those of their opponents to gain a competitive edge.
- Sports science and sports administration department faculty and students will benefit from improving their existing curriculum with a more data-centric and scientific approach.
- Sports fans and fantasy league participants will now have an opportunity to study the trend and pattern of their favorite teams or players with complete and realistic performance data.

- Information systems and IT department faculty and students will also have a unique opportunity to explore the many complex and fascinating aspects of sports and gaming applications and identify new directions for data mining research.

3 Future Directions

The full application of sports data mining is still in its infancy. While several pioneering organizations are beginning to harness their data through advanced statistical/predictive analyses, many are struggling with the prospect of adopting such systems let alone using them as a competitive advantage. As larger market professional sports organizations increase payrolls to meet demands for talent, knowledge management and data mining can be leveraged by the smaller market teams as a tool to remain competitive. This competitive balance has begun to return parity to sporting leagues knocked off-balance by payroll inequities. However, as more organizations begin to embrace these knowledge eschewing principles, it won't be long before an arms race of sorts develops, where two teams emerge; the players on the field and the analysts in the back office. Both of which will work together to propel the organization forward. Also, future advances such as distributed Artificial Intelligence that uses multiple agents or new applications of existing algorithms borrowed from the disciplines of computer science or physics, may revolutionize sports data mining. Similarly, centralized public data repositories constructed either by governments or a collective of fans, will also allow for the continuation of these techniques for teams, performance measures and predictive purposes. It will be interesting to see where the next few years will take us.

References

82games.com 2008. A Visitor's Guide to 82games.com. Retrieved Feb 20, 2008, from http://82games.com/newuser.htm.

Accuscore 2009. The Leader in Sports Forecasting. Retrieved Aug 31, 2009.

Ackoff, R. 1989. From Data to Wisdom. *Journal of Applied Systems Analysis* 16: 3–9.

Adler, J. 2006. *Baseball Hacks*. O'Reilly, Beijing.

Alavi, M. & D. E. Leidner 2001. Review: Knowledge Management and Knowledge Management Systems: Conceptual Foundations and Research Issues. *MIS Quarterly* 25(1): 107–136.

Albert, J. 1997. An Introduction to Sabermetrics. Retrieved Jan 30, 2008, from http://www-math.bgsu.edu/~albert/papers/saber.html.

Albert, J. 2008. Streaky Hitting in Baseball. *Journal of Quantitative Analysis in Sports* 4(1).

Allsopp, P. & S. Clarke 2004. Rating Teams and Analysing Outcomes in One-Day and Test Cricket. *Journal of the Royal Statistical Society: Series A* 167(4): 657–667.

Almond, E. 1994. World Cup USA '94 Unforgiveable. *The Los Angeles Times*.

Arnovitz, K. 2009. Stephen Curry, Blake Griffin, and Hasheem Thabeet: Inside the Numbers. Retrieved Aug 31, 2009,from http://myespn.go.com/blogs/truehoop/0-41-131/Stephen-Curry–Blake-Griffin–and-Hasheem-Thabeet–Inside-the-Numbers.html.

Audi, T. & A. Thompson 2007. Oddsmakers in Vegas Play New Sports Role. *The Wall Street Journal*. A1.

Babaguchi, N. & J. Ohara, et al. 2007. Learning Personal Preference from Viewer's Operations for Browsing and its Application to Baseball Video Retrieval and Summarization. *IEEE Transactions on Multimedia* 9(5): 1016–1025.

Ball, A. 2008. Winning by Numbers. *The Guardian*. London.

Ballard, C. 2005. Measure of Success. *Sports Illustrated*.

Ballard, C. 2008. Doing a Number on Soccer. *Sports Illustrated*.

Barlas, I. & A. Ginart, et al. 2005. Self-Evolution in Knowledgebases. *IEEE AutoTestCon*, Orlando, FL.

Barros, C. P. & S. Leach 2006. Performance Evaluation of the English Premier Football League with Data Envelopment Analysis. *Applied Economics* 38(12): 1449–1458.

Barry, D. 2009. Pappus' Plane – Cricket Stats. Retrieved June 6, 2009, from http://pappubahry.blogspot.com/.

Baseball Info Solutions 2003. *The Bill James Handbook*. ACTA Publications, Chicago.

Baseball-Reference.com 2008. Baseball-Reference. Retrieved Feb 20, 2008, from http://www.baseball-reference.com/.

Basketball-Reference.com 2008. Calculating PER. Retrieved Jan 30, 2008, from http://www.basketball-reference.com/about/per.html.

BBall 2008. Online Interactive Historical Sports Statics Databases. Retrieved Jan 30, 2008, from http://www.bballsports.com/.

Beech, R. 2008a. NBA Clutch Players, Part III. Retrieved Jan 30, 2008, from http://www.82games.com/clutchplay3.htm.

Beech, R. 2008b. NBA Player Shot Zones. Retrieved Jan 30, 2008, from http://www.82games.com/shotzones.htm.

Berry, S. 2005. Introduction to the Methodologies and Multiple Sports Articles. In *Anthology of Statistics in Sports*, J. Albert, J. Bennett & J. Cochran. Cambridge University Press, Alexandria, VA.

Bhandari, I. 1995. Attribute Focusing: Data Mining for the Layman. *Research Report RC 20136*. IBM TJ Watson Research Center.

Bhandari, I. & E. Colet, et al. 1997. Advanced Scout: Data Mining and Knowledge Discovery in NBA Data. *Data Mining and Knowledge Discovery* 1(1): 121–125.

Bialik, C. 2007. Tracking How Far Soccer Players Run. *The Wall Street Journal*.

Biehl, J. 2005. Is German Soccer Rigged? *Der Spiegel*. Berlin, Germany.

Bierly, P. E. & E. H. Kessler, et al. 2000. Organizational Learning, Knowledge and Wisdom. *Journal of Organizational Change Management* 13(6): 595–618.

Birnbaum, P. 2008. Sabermetrics. Retrieved Feb 16, 2008, from http://philbirnbaum.com/.

blinkx.com 2009. Blinkx Brings Users Courtside with Sports Video ootage from FoxSports.com on MSN. Retrieved Nov 4, 2009, from http://www.blinkx.com/article/blinkx-brings-users-courtside-sports-video-footage-foxsports~1031.

Boisot, M. & A. Canals 2004. Data, Information and Knowledge: Have We Got it Right? *Journal of Evolutionary Economics* 14(1): 43–67.

Brewer, P. 2009. I Don't Like Cricket. Retrieved June 6, 2009, from http://sabermetriccricket.blogspot.com.

Burns, E. & R. Enns, et al. 2006. The Effect of Simulated Censored Data on Estimates of Heritability of Longevity in the Thoroughbred Racing Industry. *Genetic Molecular Research* 5(1): 7–15.

Cameron, C. 2008. *You Bet, The Betfair Story: How Two Men Changed The World of Gambling*, HarperCollins Publishers, London, UK.

Carlisle, J. P. 2006. Escaping the Veil of Maya – Wisdom and the Organization. *39th Hawaii International Conference on System Sciences*, Koloa Kauai, HI.

Carroll, B. & P. Palmer, et al. 1998. *The Hidden Game of Football: The Next Edition*. Total Sports Inc., New York, NY.

Casino City Times 2009. Possible Gambling Scandal in Colombian Soccer. Retrieved Nov 21, 2009, from http://www.casinocitytimes.com/news/article/possible-gambling-scandal-in-colombian-soccer-178406.

Chang, C.-W. & S.-Y. Lee 1997. A Video Information System for Sport Motion Analysis. *Journal of Visual Languages and Computing* 8(3): 265–287.

Chen, H. 2001. *Knowledge Management Systems – A Text Mining Perspective*. The University of Arizona – Dept of Management Information Systems, Tucson.

Chen, H. 2006. *Intelligence and Security Informatics for International Security: Information Sharing and Data Mining*. Springer, New York, NY.

Chen, H. & M. Chau 2004. Web Mining: Machine Learning for Web Applications. *Annual Review of Information Science and Technology (ARIST)* 38: 289–329.

Chen, H. & P. Rinde, et al. 1994. Expert Prediction, Symbolic Learning, and Neural Networks: An Experiment in Greyhound Racing. *IEEE Expert* 9(6): 21–27.

Chen, H.-S. & H.-T. Chen, et al. 2007. Pitch by Pitch Extraction from Single View Baseball Video Sequences. *IEEE International Conference on Multimedia and Expo*, Beijing, China.

Chen, S.-C. & M.-L. Shyu, et al. 2005. An Enhanced Query Model for Soccer Video Retrieval Using Temporal Relationships. *International Conference on Data Engineering*, Tokyo, Japan.

Chen, Y. 2005. Information Valuation for Information Lifecycle Management. *2nd International Conference on Autonomic Computing*, Seattle, WA.

China Daily 2009. CFA Staf Probed for Soccer Scandals. *People's Daily Online*. Beijing, China.

Choo, C. W. 1996. The Knowing Organization: How Organizations Use Information to Construct Meaning, Create Knowledge, and Make Decisions. *International Journal of Information Management* 16(5): 329–340.

Chu, W.-T. & C.-W. Wang, et al. 2006. Extraction of Baseball Trajectory and Physics-Based Validation for Single-View Baseball. *IEEE International Conference on Multimedia and Expo*, Toronto, Ontario.

Cleveland, H. 1982. Information as a Resource. *The Futurist* 16(6): 34–39.

Click, J. & C. Davenport, et al. 2006. *Baseball Between the Numbers*. Basic Books, New York.

Clipta.com 2009. About Us. Retrieved Nov 4, 2009, from http://www.clipta.com/about.

Coleman, J. & A. Lynch 2001. Identifying the NCAA Tournament "Dance Card". *Interfaces* 31(3): 76–86.

Coleman, J. & A. Lynch 2008. Score Card Rankings for 2008. Retrieved Jan 30, 2008, from http://www.unf.edu/~jcoleman/score.htm.

Coleman, J. & A. Lynch 2009. Dance Card Rankings for 2009. Retrieved Sept 21, 2009, from http://www.unf.edu/~jcoleman/dance.htm.

Colquitt, L. & N. Godwin, et al. 2001. Testing Efficiency AcrossMarkets: Evidence from the NCAA Basketball Betting Market. *Journal of Business Finance and Accounting* 28(1): 231–248.

Colston, C. 2009. In Playoffs, Crunching Picks, Crunching Numbers. *USA Today*. 8C.

Cox, A. & J. Stasko 2002. SportsVis: Discovering Meaning in Sports Statistics Through Information Visualization. *IEEE Symposium on Information Visualization*, Baltimore, MD.

Craggs, T. 2004. He Stats! He Scores. *SF Weekly*. San Francisco, CA.

CricInfo 2008. Wisden. Retrieved June 19, 2008, from http://content-www.cricinfo.com/wisdenalmanack/content/current/story/almanack/.

CricketAnalysis.com 2009. Marginal Wins per Player in ODI Cricket. Retrieved June 6, 2009, from http:/cricketanalysis.com/marginal-wins-per-player-in-odi-cricket.

Data Mining Software 2009. A Breif History of Data Mining. Retrieved Sept 2, 2009, from http://www.data-mining-software.com/data_mining_history.htm.

DataSoftSystems 2009. Data Mining – History and Influences. Retrieved Sept 2, 2009, from http://www.datasoftsystem.com/articles/article-1380.html.

Davenport, T. & L. Prusak 1998. *Working Knowledge*. Harvard Business School Press, Cambridge, MA.

Delaney, S. 2006. Scandal Hits Italian Soccer on Eve of the World Cup. *The Washington Post*. Washington, DC.

Digital Scout 2008. Digital Scout. Retrieved Feb 20, 2008, from http://www.digitalscout.com/.

Ding, Y. & G. Fan 2007. Segmental Hidden Markov Models for View-based Sport Video Analysis. *IEEE Conference on Computer Vision and Pattern Recognition*, Minneapolis, MN.

Dobra, J. & T. Cargill, et al. 1990. Efficient Markets for Wagers: The Case of Professional Basketball Wagering. In *Sportometrics*, B. Goff & R. Tollison. Texas A&M University Press, College Station, TX, 215–249.

Donegan, L. 2008. Tottenham Net Role in Revolution as Beane Pitches Success on a Budget. *The Guardian*. London.

Dong, D. & R. Calvo 2007. Integrating Data Mining Processes within the Web Environment for the Sports Community. *IEEE International Conference on Integration Technology*, Shenzhen, China.

Dunshee, M. 2007. NHL Trade Deadline Revisited in Sabermetrics. Retrieved June 6, 2009, from http://www.associatedcontent.com/article/276051/nhl_trade_deadline_revisited_in_sabermetrics.html?cat=14.

Dvorak, J. & A. Junge, et al. 2000. Risk Factor Analysis for Injuries in Football Players: Possibilities for a Prevention Program. *The American Journal of Sports Medicine* 28(5): 69–74.

Fetter, H. 2003. *Taking on the Yankees – Winning and Losing in the Business of Baseball – 1903–2003*. W.W. Norton & Co., New York.

Fieltz, L. & D. Scott 2003. Prediction of Physical Performance Using Data Mining. *Research Quarterly for Exercise and Sport* 74(1): 1–25.

Flinders, K. 2002. Football Injuries are Rocket Science. *Vnunet.com*. London.

Gibbs, J. 2007. Point Shaving in the NBA: An Economic Analysis of the National Basketball Association's Point Spread Betting Market. *Dept. of Economics*. Stanford University.

Glickman, M. & H. Stern 1998. A State-Space Model for National Football League Scores. *Journal of American Statistics Association* 93: 25–35.

Goodman, A. 2005. The Market for Smart: Does Hockey Need Some PhD's? Retrieved June 6, 2009, from http://www.traffick.com/2005/06/market-for-smart-does-hockey-need-some.asp.

GRAA 2008. Greyhound Racing Association of America. Retrieved April 29, 2008, from http://www.gra-america.org.

Han, J. & M. Kamber 2001. *Data Mining: Concepts and Techniques*. Morgan Kaufmann, San Francisco.

Harville, D. 2003. The Selection or Seeding of College Basketball or Football Teams for Postseason Competition. *Journal of American Statistics Association* 98(461): 17–27.

Hirotsu, N. & M. Wright 2003. A Markov Chain Approach to Optimal Pinch Hitting Strategies in a Designated Hitter Rule Baseball Game. *Journal of Operations Research* 46(3): 353–371.

Hollinger, J. 2002. *Pro Basketball Prospectus: 2002 Edition*. Brassey's Inc., Dulles, VA.

Hopkins, W. & S. Marshall, et al. 2007. Risk Factors and Risk Statistics for Sports Injuries. *Clinical Journal of Sports Medicine* 17(3): 208–210.

Hughes, R. & E. Pfanner 2009. Raids Expose Soccer ixing Across Europe. *The New York Times*. New York, NY: A1.

Igloo Dreams 2007. Data Mining: The Referees. Retrieved Feb 6, 2008, from http://igloodreams.blogspot.com/2007/08/data-mining-referees.html.

Ilardi, S. 2007. Adjusted Plus-Minus: An Idea Whose Time Has Come. Retrieved Sept 25, 2008, from http://www.82games.com/ilardi1.htm.

Inside Edge 2008a. About Us. Retrieved Feb 20, 2008, from http://www.inside-edge.com/about_us.htm.

Inside Edge 2008b. Sample Reports. Retrieved Feb 20, 2008, from http://www.inside-edge.com/minor/sample_reports.htm.

International Association for Sports Information 2008. About IASI. Retrieved Feb 16, 2008, from http://www.iasi.org/about/index.html.

International Association on Computer Science in Sport 2008. IACSS – Objectives. Retrieved Feb 16, 2008, from http://www.iacss.org/iacss/iacss_obj.html.

James, B. 1979. *The Bill James Baseball Abstract*. Self Published.

James, B. 1982. *The Bill James Baseball Abstract*. Ballantine Books, New York.

James, B. & J. Henzler 2002. *Win Shares*. STATS Publishing, Morton Grove, IL.

Jimmy, D. 2007. Point-Shaving Ref: NBA Betting Scandal Borne of Hypocritical Anti-Betting Stance. Retrieved Feb 22, 2008, from http://www.jimmydsports.com/fantasy-sports-columns/nba-point-shaving-ref-july242007.aspx.

Johansson, U. & C. Sonstrod 2003. Neural Networks Mine for Gold at the Greyhound Track. *International Joint Conference on Neural Networks*, Portland, OR.

Justice, R. 2006. Rockets See Morey as Extra Edge. *Houston Chronicle*. Houston.

Kelley, D. & J. Mureika, et al. 2006. Predicting Baseball Home Run Records Using Exponential Frequency Distributions. Retrieved Jan 15, 2008, from http://arxiv.org/abs/physics/0608228v1.

Koning, R. 2000. Balance in Competition in Dutch Soccer. *The Statistician* 49: 419–431.

Kuper, S. 2006. *Soccer Against the Enemy*. Nation Books, New York.

Lahti, R. & M. Beyerlein 2000. Knowledge Transfer and Management Consulting: A Look at the Firm. *Business Horizons* 43(1): 65–74.

Lavigne, P. 2009. Fans Behaving Badly? Never Fear. Retrieved Nov 18, 2009, from http://sports.espn.go.com/espn/otl/news/story?id=4603176.

Lee, C. 1997. An Empirical Study of Boxing Match Prediction Using a Logistic Regression Analysis. *Section Statistics Sports, American Statistical Association, Joint Statistical Meeting*, Anaheim, CA.

Levin, R. & G. Mitchell, et al. 2000. The Report of the Independent Members of the Commissioner's Blue Ribbon Panel on Baseball Economics. Major League Baseball.

Lewis, M. 2003. *Moneyball.* W.W. Norton & Company, New York.

Liu, G. & X. Tang, et al. 2009. A Novel Approach for Tracking High Speed Skaters using a Panning Camera. *Pattern Recognition* 42(11): 2922–2935.

Lyons, K. 2005. Data Mining and Knowledge Discovery. *Australian Sports Commission Journals* 2(4).

Marakas, G. 2003. *Modern Data Warehousing, Mining, and Visualization: Core Concepts.* Prentice Hall, Upper Saddle River, NJ.

Match Analysis 2009. Video Editing, Data Collection, and Statistics for Soccer (Football). Retrieved Aug 31, 2009.

McCarthy, M. & A. Perez 2009. Pro Leagues, NCAA Resist Push by States to Legalize Betting. *USA Today.*

Merron, J. 2006. Biggest Sports Gambling Scandals. Retrieved Nov 21, 2009, from http://sports.espn.go.com/espn/page2/story?page=merron/060207.

MIT Sloan Alumni Profile 2008. Daryl Morey, MBA '00. Retrieved Jan 30, 2008, from http://mitsloan.mit.edu/mba/alumni/morey.php.

Moore, G. 2009. Bluefin Lab's Software to Scan Sports Video. Retrieved Nov 4, 2009, from http://www.masshightech.com/stories/2009/10/05/daily15-Bluefin-Labs-software-to-scan-sports-video.html.

Murphy, C. 2005. *Competitive Intelligence: Gathering, Analysing and Putting it to Work.* Gower, Burlington, VT.

Oliver, D. 2005. *Basketball on Paper: Rules and Tools for Performance Analysis.* Brassey's Inc., Dulles, VA.

Oorlog, D. 1995. Serial Correlation in the Wagering Market for Professional Basketball. *Quarterly Journal of Business and Economics* 34(2): 96–109.

O'Reilly, N. & P. Knight 2007. Knowledge Management Best Practices in National Sport Organizations. *International Journal of Sport Management and Marketing* 2(3): 264–280.

Ortiz, J. L. 2007. MLB's Online Venture is Big Hit. *USA Today.*

Page, G. 2005. Using Box Scores to Determine a Position's Contribution to Winning Basketball Games. *Dept of Statistics.* Brigham Young University.

Papahristoulou, C. 2006. Team Performance in UEFA Champions League 2005–2006. MRPA Paper No. 138.

Paul, R. & A. Weinbach 2005. Bettor Misconceptions in the NBA: The Overbetting of Large Favorites and the Hot Hand. *Journal of Sports Economics* 6(4): 390–400.

Pelton, D. 2005. The Sonics Play Moneyball: Part One. Retrieved Jan 30, 2008, from http://www.nba.com/sonics/news/moneyball050119.html.

Petro, N. 2001. Digital Scout to Provide Statistical Analysis for USA Baseball Tournament. Retrieved Feb 20, 2008, from http://www.digitalscout.com/news/news_tournament.php.

Petro, N. 2003. Digital Scout Signs Two-Year Agreement with Little League Baseball. Retrieved Feb 20, 2008, from http://www.digitalscout.com/news/news_littleleague.php.

Philpott, A. & S. Henderson, et al. 2004. A Simulation Model for Predicting Yacht Match Race Outcomes. *Operations Research* 52(1): 1–16.

Piatetsky-Shapiro, G. 2008. Difference between Data Mining and Statistics. Retrieved Oct 2, 2008, from http://www.kdnuggets.com/faq/difference-data-mining-statistics.html.

Professional Football Researchers Association 2008. Welcome!! Retrieved Feb 16, 2008, from http://www.profootballresearchers.org/index.htm.

Pro-Football-Reference.com 2008. Pro-Football-Reference. Retrieved Feb 20, 2008, from http://www.pro-football-reference.com/.

Roach, M. & J. Mason, et al. 2001. Video Genre Classification Using Dynamics. *IEEE International Conference on Acoustics, Speech, and Signal Processing*, Salt Lake City, UT.

Roach, M. & J. Mason, et al. 2002. Recent Trends in Video Analysis: A Taxonomy of Video Classification Problems. *International Conference on Internet and Multimedia Systems and Applications*, Kauai, Hawaii.

Rosenbaum, D. 2004. Measuring How NBA Players Help Their Teams. Retrieved Jan 30, 2008, from http://www.82games.com/comm30.htm.

Rosenbaum, D. 2005. A Statistical Holy Grail: The Search for the Winner Within. *The New York Times.*

Rotshtein, A. & M. Posner, et al. 2005. Football Predictions Based on a Fuzzy Model with Genetic and Neural Tuning. *Cybernetics and Systems Analysis* 41(4): 619–630.

Rue, H. & O. Salvensen 2000. Prediction and Retrospective Analysis of Soccer Matches in a League. *The Statistician* 49: 399–418.

Sandoval, G. 2006. A Video Slam Dunk for the NBA. Retrieved Jan 30, 2008, from http://www.news.com/A-video-slam-dunk-for-the-NBA/2100-1008_3-6034908.html.

SAS 2005. A Method to March Madness. Retrieved Jan 30, 2008, from http://www.sas.com/news/feature/01mar05/dancecard.html.

Schatz, A. 2006. *Pro Football Prospectus 2006: Statistics, Analysis, and Insight for the Information Age*. Workman Publishing Company.

Schell, M. J. 1999. *Baseball's All-Time Best Hitters: How Statistics Can Level the Playing Field.* Princeton University Press, Princeton, NJ.

Schumaker, R. P. 2007. Using SVM Regression to Predict Greyhound Races. *Information Systems Dept. Research Seminar*, New Rochelle, NY.

Schumaker, R. P. & H. Chen 2008. Evaluating a News-Aware Quantitative Trader: The Effects of Momentum and Contrarian Stock Selection Strategies. *Journal of the American Society for Information Science* 59(1): 1–9.

Seder, J. & C. Vickery 2005. The Relationship of Subsequent Racing Performance to Foreleg Flight Patterns During Race Speed Workouts of Unraced 2-Yr-Old Thoroughbred Racehorses at Auctions. *Journal of Equine Veterinary Science* 25(12): 505–522.

Serenko, A. & N. Bontis 2004. Meta-review of Knowledge Management and Intellectual Capital Literature: Citation Impact and Research Productivity Rankings. *Knowledge and Process Management* 11(3): 185–198.

Shilling, D. 2005. Hockey Project Rating. Retrieved June 6, 2009, from http://web.archive.org/web/20051222110731/http://members.shaw.ca/hbtn/player_study/hpr.htm.

Shulman, K. 1996. Data Mining in the Backcourt: Advanced Scout Gives Coaches an Assist. Retrieved Feb 6, 2008, from http://www.dciexpo.com/news/archives/scout.htm.

Sinins, L. 2007. Complete Baseball Encyclopedia: Version 8.0.

Smeulders, A. & M. Worring, et al. 2000. Content-Based Image Retrieval at the End of the Early Years. *IEEE Transactions on Pattern Analysis and Machine Intelligence* 22(12): 1349–1380.

Smith, L. & B. Lipscomb, et al. 2007. Data Mining in Sports: Predicting Cy Young Award Winners. *Journal of Computing Sciences in Colleges* 22(4): 115–121.

Society for American Baseball Research 2008. About SABR. Retrieved Feb 16, 2008, from http://www.sabr.org/sabr.cfm?a=cms,c,110,39,156.

Solieman, O. 2006. Data Mining in Sports: A Research Overview. *Dept. of Management Information Systems*. The University of Arizona. Tucson.

SportsVHL.com 2009. SportsVHL – Show What Ya Got! Retrieved Nov 4, 2009, from http://www.sportsvhl.com/.

Stefani, R. 1999. A Taxonomy of Sports Rating Systems. *IEEE Transactions on Systems, Man, and Cybernetics – Part A* 29(1): 116–120.

Stern, H. 1991. On Probability of Winning a Football Game. *Journal of American Statistics Association* 45: 179–183.

Stern, H. 2004. Statistics and the College Football Championship. *The American Statistician* 58(3): 179–185.

Taggart, S. 1998. Olympic Data Crunching. Retrieved Feb 6, 2008, from http://www.wired.com/science/discoveries/news/1998/08/14175.

Takagi, S. & S. Hattori, et al. 2003. Sports Video Categorizing Method Using Camera Motion Parameters. *International Conference on Multimedia and Expo*, Baltimore, MD.

The Association for Professional Basketball Research 2008. APBR.org. Retrieved Feb 16, 2008, from http://apbr.org.

The New York Times Staff 2004. *The New York Times Guide to Essential Knowledge: A Desk Reference for the Curious Mind*. St. Martin's Press, New York.

Thomas, A. 2006. The Impact of Puck Possession and Location on Ice Hockey Strategy. *Journal of Quantitative Analysis in Sports* 2(1).

Thorn, J. & P. Palmer 1984. *The Hidden Game of Baseball*. Doubleday, Garden City, NJ.

TRACAB 2007. TRACAB in Champions League. Retrieved April 4, 2008, from http://www.tracab.com/news.asp?id=25.

Truong, B. T. & C. Dorai 2000. Automatic Genre Identification for Content-Based Video Categorization. *International Conference on Pattern Recognition*, Barcelona, Spain.

truveo.com 2009. Truveo Video Search. Retrieved Nov 4, 2009, from http://www.truveo.com/category/Sports#category%3A%22sports%22.

Tversky, A. & T. Gilovich 2004. The Cold Facts About the "Hot Hand" in Basketball. In *Preference, Belief, and Similarity: Selected Writings*, A. Tversky & E. Shafir. MIT Press, Cambridge, MA.

USA Today 2008. Jeff Sagarin NCAA Basketball Ratings. Retrieved Sept 27, 2008, from http://www.usatoday.com/sports/sagarin/bkt0708.htm.

Voigt, D. 1969. America's First Red Scare – The Cincinnati Reds of 1869. *Ohio History* 78: 13–24.

WasWatching.com 2005. Stat Glossary Archives. Retrieved Sept 21, 2009.

Weeks, C. 2006. Digital Scout Software Powers the Stats for the Inaugural Arizona Cactus Classic. Retrieved Feb 20, 2008, from http://www.digitalscout.com/news/news_az_cactus_classic_06.php.

White, P. 2006. Scouts Uncover a Winning Edge. *USA Today*. New York: 7E.

Willoughby, K. 1997. Determinants of Success in the CFL: A Logistic Regression Analysis. *National Annual Meeting to the Decision Sciences Institute*, Atlanta, GA.

Witten, I. H. & Frank, E. 2005. "Data Mining: Practical machine learning tools and techniques, 2nd Edition". Morgan Kaufmann, San Francisco. http://www.cs.waikato.ac.nz/~ml/weka/book.html.

Wolfers, J. 2006. Point Shaving: Corruption in NCAA Basketball. *AEA Papers and Proceedings* 96(2): 279–283.

Woolner, K. 2006. Why is Mario Mendoza So Important? In *Baseball Between the Numbers*, J. Keri. Basic Books, New York.

Xinhua News 2009. Ukraine's Dynamo, Shakhtar among world's top 10 soccer clubs. *Xinhua*. Beijing.

Yang, T. Y. & T. Swartz 2004. A Two-Stage Bayesian Model for Predicting Winners in Major League Baseball. *Journal of Data Science* 2(1): 61–73.

Yow, D. & B.-L. Yeo, et al. 1995. Analysis and Presentation of Soccer Highlights from Digital Video. *Asian Conference on Computer Vision*, Singapore.

Zeleny, M. 1987. Management Support Systems: Towards Integrated Knowledge Management. *Human Systems Management* 7(1): 59–70.

Zhu, B. & H. Chen 2005. Information Visualization. *Annual Review of Information Science and Technology (ARIST)* 39: 139–178.

Zimmerman, P. 1985. *The New Thinking Man's Guide to Pro Football*. Simon and Schuster, New York.

Index

C

MIX
Papier aus verantwortungsvollen Quellen
Paper from responsible sources
FSC® C105338

If you have any concerns about our products,
you can contact us on
ProductSafety@springernature.com

In case Publisher is established outside the EU,
the EU authorized representative is:
Springer Nature Customer Service Center GmbH
Europaplatz 3, 69115 Heidelberg, Germany

Printed by Libri Plureos GmbH
in Hamburg, Germany